Contemporary's

NUMBER POWER

Level D

Intermediate 2

Wright Group

ISBN: 0-8092-0611-0

Send all inquiries to:
Wright Group/McGraw-Hill
130 E. Randolph, Suite 400
Chicago, IL 60601

Printed in the United States of America.

14 15 16 17 REL 12 11 10

The McGraw·Hill Companies

Contents

To the Learner

Even if math has never been easy for you, this text will give you the instruction and practice you need to understand the basics. In this global and technological society, an understanding of math is important, and at some time or another, you will be asked to demonstrate that you can solve math problems well.

Using *Number Power Intermediate 2, Level D* is a good way to develop and improve your mathematical skills. It is a comprehensive text for mathematical instruction and practice. Beginning with basic comprehension skills in addition, subtraction, multiplication, and division, *Number Power Intermediate 2, Level D* includes basic concepts about performing basic computations. For instance, concepts include information such as words like *plus, sum,* and *total* often signal addition problems, the addition symbol is a "+" sign, and addition problems have distinctive characteristics. Then problems, which illustrate the basic kinds of addition problems learners will confront, are included for plenty of practice.

Accompanying all of this practice are a Skills Inventory Pre-Test and a Skills Inventory Post-Test. The Skills Inventory Pre-Test will help you identify your math strengths and weaknesses before you begin working in the book. Then you can work in those areas where additional instruction and practice are needed. Upon completion of these exercises, you should take the Skills Inventory Post-Test to see if you have achieved mastery. Mastery is whatever score you and your instructor have agreed upon to be correct to insure that you understand each group of problems.

Usually mastery is completing about 80 percent of the problems correctly. After achieving mastery, you should then move to the next section of instruction and practice. In this way the text offers you the chance to learn at your own pace, covering only the material that you need to learn. In addition, the instruction on each page offers you the opportunity to work on your own.

Addition, subtraction, multiplication, division, decimals, signed numbers, ratio and percent instruction and practice are presented in the first part of *Number Power Intermediate 2, Level D.* The other part contains applications. It includes numeration, number theory, data interpretation, algebra, measurement, and geometry. You will also work with ordinal numbers, place value of numbers, graphs, tables, charts, number sentences, calendars, time, plane figures, logical reasoning, problem solving, estimation, and many other topics.

Completing *Number Power Intermediate 2, Level D* will make you more confident about doing mathematical problems. Remember to use the Answer Key in the back of the book to check your responses. Soon you will find yourself either enjoying math for the first time or liking it even more than you did previously.

Skills Inventory Pre-Test

Part A: Computation

Circle the letter for the correct answer for each problem.

1

$$\begin{array}{r} 16 \\ \times\ 8 \\ \hline \end{array}$$

A 328 **C** 88
B 200 **D** 128
E None of these

2

$3\overline{)78}$

F 36
G 16
H 22 r 2
J 22
K None of these

3

$2.45 − $1.96 = _____

A $0.49
B $1.49
C $0.59
D $0.48
E None of these

4

$$\begin{array}{r} 206 \\ \times\ 12 \\ \hline \end{array}$$

F 620
G 2,492
H 2,472
J 2,462
K None of these

5

$\frac{3}{15} + \frac{7}{15} =$ _____

A $\frac{2}{3}$ **C** $\frac{2}{5}$

B $\frac{1}{3}$ **D** $\frac{1}{10}$

E None of these

6

25% of 160 = _____

F 64 **H** 40
G 25 **J** 400
K None of these

7

−10 + 2 = _____

A 8
B −8
C 12
D −12
E None of these

8

6.37 + 0.043 = _____

F 6.8
G 0.6413
H 6.3743
J 6.413
K None of these

9

$$\begin{array}{r} 3\frac{1}{5} \\ -1\frac{1}{10} \\ \hline \end{array}$$

A $2\frac{1}{10}$ **C** $2\frac{1}{5}$

B $2\frac{1}{15}$ **D** $2\frac{9}{10}$

E None of these

10

0.2503 × 0.5 = _____

F 1.2515
G 0.012515
H 0.12515
J 12.51
K None of these

11

$13\overline{)273}$

A 20 r 3
B 19 r 25
C 21
D 19
E None of these

12

−10 × (−5) = _____

F 50
G −50
H 15
J 2
K None of these

13

40% of ☐ = 4

A 1
B 10
C 100
D 1000
E None of these

14

What percent of 80 is 20?

F 40%
G 4%
H 25%
J 20%
K None of these

Part B: Applied Mathematics

Circle the letter for the correct answer to each problem.

Every year since 1993, scientists have asked American teens if they had smoked in the last 30 days. This graph shows the percentage of teens that said yes. Study the graph. Then use it to do Numbers 15 through 18.

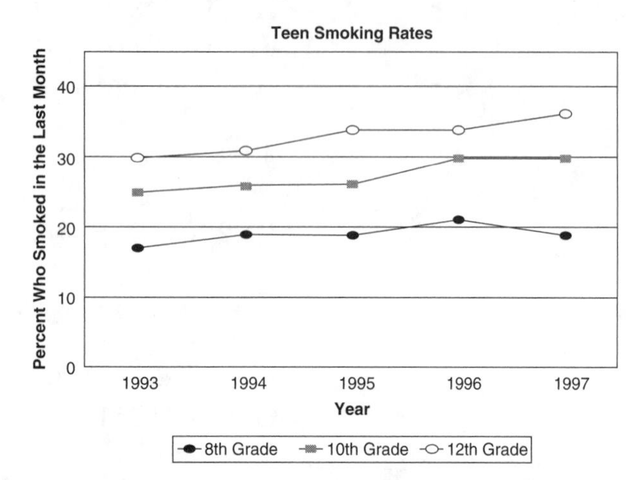

15 By about how much did the 12th-grade smoking rate rise between 1993 and 1997?

A 10 percent
B 5 percent
C 15 percent
D 35 percent

16 Which of the following best describes how the 8th-grade smoking rate is changing?

F It is rising steadily, without pause.
G It never changes.
H It has started to decline.
J It has dropped drastically.

17 In what period was there the greatest increase in the percentage of 10th graders who smoked?

A 1993 to 1994
B 1994 to 1995
C 1995 to 1996
D 1996 to 1997

18 Which of the following statements is supported by the information on this graph?

F As teenagers get older, they are more likely to smoke.
G The older you are, the harder it is to stop smoking.
H The older teenage smokers get, the more cigarettes they smoke per day.
J The older teenagers get, the less likely they are to start smoking.

---

19 If you list all the odd numbers, which of the following numbers will *not* be on your list?

A 49
B 66
C 77
D 25

20 What is the measure of $\angle ACB$ in this triangle?

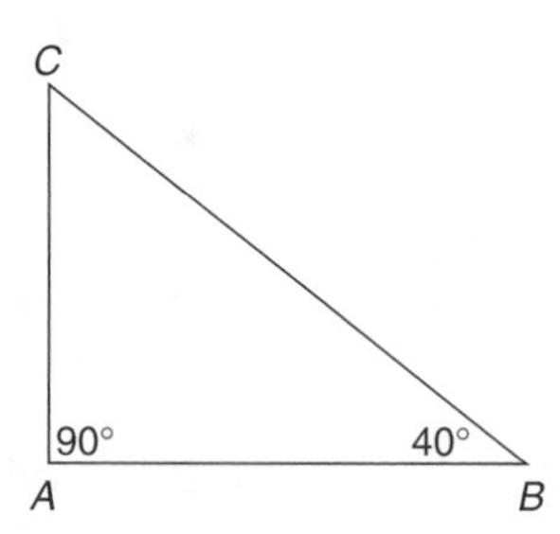

F 45°
G 40°
H 50°
J This cannot be determined.

Use this chart for Numbers 21–24.

Fried Rice Mix Nutrition Facts

Serving Size:	$\frac{1}{3}$ cup (44 grams)	
Servings per Container	5	
Amount Per Serving		
Calories	150	
Calories from Fat	10	
		% Daily Allowance
Total Fat	1 g	2%
Saturated Fat	0 g	0%
Cholesterol	0 mg	0%
Sodium	600 mg	25%
Potassium	420 mg	12%
Total Carbohydrates	32 g	11%
Dietary Fiber	2 g	9%
Sugars	6 g	
Other Carbohydrates	24 g	
Protein	5 g	

21 This chart shows that 600 mg (milligrams) of sodium is 25% of the sodium you should eat each day. What is the recommended daily allowance of sodium?

A 1,200 mg **C** 2,400 mg
B 150 mg **D** 15,000 mg

22 How many cups of fried rice mix are in this container?

F 5 **H** $\frac{3}{5}$
G $1\frac{2}{3}$ **J** 3

23 About how many ounces of fried rice mix are in one serving? (1 ounce = 28 grams)

A 1.5 **C** 1,200
B 15 **D** 4

24 Which number sentence shows how many calories are in 1 cup of this mix?

F $150 \div 3 = \square$
G $150 \times 3 = \square$
H $3 \div 150 = \square$
J $150 = \square$

An end table is on sale for $123.00. To buy it, you must put 15% down. **Use this information to do Numbers 25 through 27.**

25 The table has been marked down 20%. Which of these is the best estimate of how much the table originally cost?

A $100.00
B $150.00
C $175.00
D $200.00

26 Which of these number sentences could you use to figure out how much is left to pay after you make the down payment?

F $\$123.00 \times 0.15 = \_\_\_\_$
G $\frac{\$123.00}{6} = \_\_\_\_$
H $\$123.00 - (\$123. \times 0.15) = \_\_\_\_$
J $\$123.00 - 0.15 = \_\_\_\_$

27 Mark and Cindy have $600.00 to buy furniture. If they buy an end table and a chair for $345.59, about how much will they have left? *(Round all numbers to the nearest 10 dollars.)*

A $250.00
B $130.00
C $70.00
D $160.00

28 Which of these has the same value when it is rounded to the nearest tenth or to the nearest hundredth?

F 1.452 **H** 1.501
G 1.512 **J** 1.507

When you travel to other countries, you must trade some of your U.S. dollars for the local currency. This chart shows what you can get in different countries for one U.S. dollar. Study the chart. Then do Numbers 29 through 33.

Value of One U.S. Dollar

1.36	Canadian dollars
5.12	French francs
1.5	German deutsche marks
35.43	Indian rupees
1,543	Italian lira
108.78	Japanese yen
7.6	Mexican pesos
1.56	English pounds

29 What is the smallest exchange rate (the smallest number) shown in the chart?

A 1.36 **C** 108.78
B 1.5 **D** 1.56

30 Which of the following is another way to show how many German deutsche marks you can get in exchange for one U.S. dollar?

F $1\frac{1}{5}$ deutsche marks for each U.S. dollar

G $1\frac{1}{25}$ deutsche marks for each U.S. dollar

H $1\frac{1}{2}$ deutsche marks for each U.S. dollar

J $1\frac{1}{50}$ deutsche marks for each U.S. dollar

31 How many Mexican pesos should you get in exchange for 25 U.S. dollars?

A 3 **C** 1,900
B 190 **D** 300

32 You find an Italian sweater you like. It is marked 80,000 lira, or 50 U.S. dollars. How many lira would you save if you bought the sweater with dollars?

F 2,850
G 7,150
H 77,150
J This cannot be determined.

33 Which of these is the value of a U.S. dollar in yen?

A one hundred eighty and seventy-eight hundredths
B one hundred eighty and seventy-eight thousandths
C one hundred eight and seventy-eight tenths
D one hundred eight and seventy-eight hundredths

---

34 What number goes in the box to make the number sentence true?

$5 \times 7 \times \square = 0$

F 35
G –35
H 0
J 1

This diagram is part of the plans for making a simple bank. Study the diagram. Then do Numbers 35 through 38.

Coin Bank

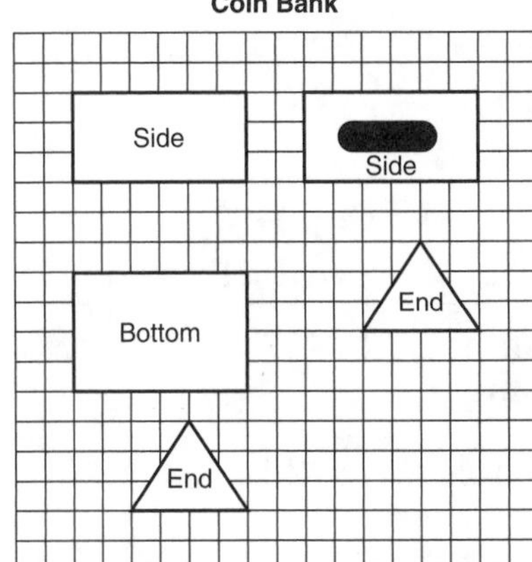

Each grid square is 1 inch by 1 inch.

35 Which two pieces, if any, are congruent?

A the sides
B the ends
C the bottom and the blank side
D None

36 What area will the bottom of this bank cover?

F 10 square inches
G 24 square inches
H 20 square inches
J This cannot be determined from the diagram.

37 What type of triangle does each end form?

A equilateral
B obtuse
C right
D isosceles

38 Roger made one of these banks. It took 55 minutes to cut out all the pieces, 23 minutes to put them together, and 32 minutes to sand and finish the bank. Which number sentence could Roger use to estimate how many hours he spent on the bank?

F $\frac{60 + 20 + 30}{60} = \_\_\_\_$

G $60(60 + 20 + 30) = \_\_\_\_$

H $(60 + 20 + 30) - 60 = \_\_\_\_$

J $\frac{60 + 20 + 30}{12} = \_\_\_\_$

Skills Inventory Pre-Test Evaluation Chart

Use the key to check your answers on the Skills Inventory Pre-Test. The Evaluation Chart shows where to turn in the book to find help with the problems you missed.

Key

1	D
2	K
3	A
4	H
5	A
6	H
7	B
8	J
9	A
10	H
11	C
12	F
13	B
14	H
15	B
16	H
17	C
18	F
19	B
20	H
21	C
22	G
23	A
24	G
25	B
26	H
27	A
28	H
29	A
30	H
31	B
32	F
33	D
34	H
35	B
36	G
37	D
38	F

Evaluation Chart

Problem	Skill	For Help, Study Pages
1,4	Multiplication of Whole Numbers	14–17
2,11	Division of Whole Numbers	18–22
3, 8, 10	Decimals	30–41
5, 9	Fractions	42–59
7, 12	Integers	60–66
6, 13, 14	Ratios, Proportions, Percents	67–79
29, 30, 33	Numeration, Number Theory	1–5, 30–31, 42–47, 60
15, 16, 17, 18	Data Interpretation	80–93
19, 24, 26, 34	Pre-Algebra, Algebra	94–105
23, 36, 38	Measurement	106–121
20, 35, 37	Geometry	122–136
21, 22, 31, 32	Computation in Context	23–26, 39, 57, 64
25, 27, 28	Estimation, Rounding	6–7, 27

Number Power Intermediate 2, Level D

Correlations Between Number Power Intermediate 2, Level D *and TABE™ Mathematics Computation*

Multiplication of Whole Numbers **Pre-Test Score** ☐ **Post-Test Score** ☐

Subskill	TABE, Form 7	TABE, Form 8	Practice and Instruction Pages in This Text (*p* means practice page.)	Additional Practice and Instruction Resources
Multiplying by 1 digit with regrouping	5	1, 3	14, 16, 28–29*p*	*Number Sense,* Bk. 2, pages 1–12 *Breakthroughs in Math / Bk. 1,* pages 72–76, 84–87 *Number Power, Bk. 1,* pages 45–49, 54–55
Multiplying by 2 or more digits with regrouping	3, 6, 7	8, 10	15–17, 28–29*p*	*Number Sense,* Bk. 2, pages 13–18 *Breakthroughs in Math / Bk. 1,* pages 77–81, 88–91 *Number Power, Bk. 1,* pages 50–59

Division of Whole Numbers **Pre-Test Score** ☐ **Post-Test Score** ☐

Subskill	TABE, Form 7	TABE, Form 8	Practice and Instruction Pages in This Text	Additional Practice and Instruction Resources
Dividing by 1 digit with no remainder	1, 4, 9	2, 4	18, 20–21, 28–29*p*	*Number Sense,* Bk. 2, pages 23–26 *Breakthroughs in Math / Bk. 1,* pages 100–107, 110–112 *Number Power, Bk. 1,* pages 72–75, 89
Dividing by 2 or more digits with no remainder	8	7	22, 28–29*p*	*Breakthroughs in Math / Bk. 1,* pages 120–121 *Number Power, Bk. 1,* pages 80–82, 84, 86, 89
Dividing by 1 digit with a remainder		9	19–21, 28–29*p*	*Number Sense,* Bk. 2, pages 27–33 *Breakthroughs in Math / Bk. 1,* pages 108–112, 116–117 *Number Power, Bk. 1,* pages 76–79, 89

Corresponds to TABE™ Forms 7 and 8
Tests of Adult Basic Education are published by CTB Macmillan/McGraw-Hill.
Such company has neither endorsed nor authorized this test preparation book.

Decimals **Pre-Test Score** ☐ **Post-Test Score** ☐

Subskill	TABE, Form 7	TABE, Form 8	Practice and Instruction Pages in This Text	Additional Practice and Instruction Resources
Adding decimals	2	6, 12	32, 40–41*p*	*Number Sense,* Bk. 3, pages 19–28 *Breakthroughs in Math / Bk. 2,* pages 42–43 *Pre-GED Satellite Program: Mathematics,* pages 53–56
Subtracting decimals	12, 13	5	33, 40–41*p*	*Number Sense,* Bk. 3, pages 33–41, 44 *Breakthroughs in Math / Bk. 2,* pages 44–45 *Pre-GED Satellite Program: Mathematics,* pages 53, 54, 56
Multiplying decimals	17	22	34–35, 40–41*p*	*Number Sense,* Bk. 4, pages 1–17 *Breakthroughs in Math / Bk. 2,* pages 48–51 *Pre-GED Satellite Program: Mathematics,* pages 57–60
Dividing with decimals	11	24	36–38, 40–41*p*	*Number Sense,* Bk. 4, pages 22–37 *Breakthroughs in Math / Bk. 2,* pages 52–57 *Pre-GED Satellite Program: Mathematics,* pages 61–63

Fractions **Pre-Test Score** ☐ **Post-Test Score** ☐

Subskill	**TABE, Form 7**	**TABE, Form 8**	**Practice and Instruction Pages in This Text**	**Additional Practice and Instruction Resources**
Adding fractions		18, 19	48, 51–52, 58–59*p*	*Number Sense,* Bk. 6, pages 1–17, 20–24 *Breakthroughs in Math/ Bk. 2,* pages 66–70, 73–75, 104–105 *Number Power, Bk. 2,* pages 7–11, 13–19
Subtracting fractions	10, 14	16	49–52, 58–59*p*	*Number Sense,* Bk. 6, pages 27–30 *Breakthroughs in Math/ Bk. 2,* pages 66–69, 76–78 *Number Power, Bk. 2,* pages 7–9, 11, 21
Multiplying fractions		13	53–54, 58–59*p*	*Number Sense,* Bk. 7, pages 1–14 *Breakthroughs in Math/ Bk. 2,* pages 66–69, 92–94 *Number Power, Bk. 2,* pages 7–9, 11, 28–31
Dividing with fractions	15, 23		55–56, 58–59*p*	*Number Sense,* Bk. 7, pages 26–31, 37 *Breakthroughs in Math/ Bk. 2,* pages 66–69, 96–98 *Number Power, Bk. 2,* pages 7–9, 11, 34–39

Integers **Pre-Test Score** ☐ **Post-Test Score** ☐

Subskill	**TABE, Form 7**	**TABE, Form 8**	**Practice and Instruction Pages in This Text**	**Additional Practice and Instruction Resources**
Adding integers	22	20	61, 64–66*p*	*Real Numbers,* Bk. 5, pages 30–31, 33–36, 41, 49 *Number Power,* Bk. 3, pages 4–11, 18 *GED Satellite Program, Mathematics,* pages 279–283, 289
Subtracting integers	18, 25	21, 25	62, 64–66*p*	*Real Numbers,* Bk. 5, pages 37–41, 49 *Number Power,* Bk. 3, pages 12–13, 18 *GED Satellite Program, Mathematics,* pages 283–285, 289
Dividing with integers	20	11	63, 64–66*p*	*Real Numbers,* Bk. 5, pages 46–49 *Number Power,* Bk. 3, pages 16–17 *GED Satellite Program, Mathematics,* pages 287–289

Percent **Pre-Test Score** ☐ **Post-Test Score** ☐

Subskill	**TABE, Form 7**	**TABE, Form 8**	**Practice and Instruction Pages in This Text**	**Additional Practice and Instruction Resources**
Percent	16, 19, 21, 24	14, 15, 17, 23	72–79*p*	*Number Power,* Bk. 2, pages 76–81, 85–86, 89–90 *Breakthroughs in Math / Bk. 2,* pages 114–122, 124–126, 132, 136, 141 *The GED Math Problem Solver,* pages 176–180, 182–183

Correlations Between Number Power Intermediate *2, Level D and TABE Applied Mathematics*

Numeration **Pre-Test Score** ☐ **Post-Test Score** ☐

Subskill	TABE, Form 7	TABE, Form 8	Practice and Instruction Pages in This Text	Additional Practice and Instruction Resources
Word names	31		3, 5, 8–9*p*, 30, 40–41*p*	*Number Sense,* Bk. 1, pages 1–6; Bk. 3 pages 9, 13–16; Bk. 9, pages 1–11 *Breakthroughs in Math / Bk. 1,* pages 7–11; *Bk. 2,* pages 34–38, 114–115 *Number Power Review,* pages 2–3, 16–17, 44–47, 60–61
Recognizing numbers	1	1	3, 5, 8–9*p*, 30, 40–41*p*, 60	*Number Sense,* Bk. 1, pages 1–6; Bk. 3, pages 9, 13–16; Bk. 9, pages 1–11 *Breakthroughs in Math / Bk. 1,* pages 7–11; *Bk. 2,* pages 34–38, 114–115 *Number Power Review,* pages 2–3, 16–17, 44–47, 60–61
Expanded notation	6		1–2, 8–9*p*, 30	*Number Sense,* Bk. 1, pages 1–4, 7 *Number Sense,* Bk. 3, pages 7–11, 13, 14, 16
Comparison	8	5, 17	4, 8–9*p*, 31, 40–41*p*, 43, 46, 58–59*p*, 81, 85, 88	*Number Sense,* Bk. 1, pages 1–4, 8–9, 54–55; Bk. 3, pages 1–6, 17–18; Bk. 5, pages 14–17, 25–27 *Breakthroughs in Math / Bk. 2,* pages 39, 71, 106–107 *Number Power Review,* pages 2–4, 16–17, 48, 51–63
Fractional part	13	19, 32, 45	42–47, 58–59*p*, 81, 85–86	*Number Sense,* Bk. 5, pages 1–13, 18–60; Bk. 9, pages 12, 13, 16–32 *Breakthroughs in Math / Bk. 2,* pages 66–69, 72, 104–106, 117, 120–122 *Number Power,* Bk. 2 pages 5–9, 52–53, 78–79
Number line	3		60–62	*Number Sense,* Bk. 3, pages 1, 2; Bk. 5, pages 26–27 *Pre-GED Satellite Program: Mathematics,* pages 50–51, 81–82, 101–103 *The GED Math Problem Solver,* pages 8, 14, 15, 36–39, 119, 126–131

Number Theory **Pre-Test Score** ☐ **Post-Test Score** ☐

Subskill	**TABE, Form 7**	**TABE, Form 8**	**Practice and Instruction Pages in This Text**	**Additional Practice and Instruction Resources**
Equivalent form		30	5, 8–9*p*, 44, 46–47, 51, 58–59*p*	*Breakthroughs in Math / Bk. 2,* pages 68–70, 104–105, 116–122 *Number Power,* Bk. 2, pages 8–12, 49, 52–53, 72–80 *Pre-GED Satellite Program: Mathematics,* pages 69–79, 161–167
Ratio, Proportion		26, 27	67–71, 78–79*p*	*Number Sense,* Bk. 8, page 1–60 *Pre-GED Satellite Program: Mathematics,* pages 149–160, 175–179 *The GED Math Problem Solver,* pages 160–175
Percent	14, 15, 25, 34, 43	9, 14, 44	72–79*p*	*Number Power,* Bk. 2, page 81–92 *Breakthroughs in Math / Bk. 2,* pages 114–141 *The GED Math Problem Solver,* pages 176–183

Data Interpretation **Pre-Test Score** ☐ **Post-Test Score** ☐

Subskill	**TABE, Form 7**	**TABE, Form 8**	**Practice and Instruction Pages in This Text**	**Additional Practice and Instruction Resources**
Graphs	10, 11, 36, 47, 48, 49	7, 37, 38, 39	84–93*p*	*Number Power,* Bk. 8, pages 1–32, 38–54, 72–82, 96–99, 136–144, 156–159 *Real Numbers,* Bk. 3, pages 12–48, 61–66 *GED Satellite Program, Mathematics,* pages 177–187
Probability, Statistics	35, 50		83, 92–93*p*	*Number Power,* Bk. 8, pages 60–67, 83–88 *Real Numbers,* Bk. 3, pages 51–60 *Pre-GED Satellite Program, Mathematics,* pages 185–190
Tables, charts, and diagrams		10	80–82, 92–93*p*	*Number Power,* Bk. 5, pages 67–91 *Real Numbers,* Bk. 3, pages 1–16, 44–50 *GED Satellite Program, Mathematics,* pages 188–191
Pre-Solution		8	23–27, 40*p*, 71	*Number Power,* Bk. 7, pages 2–69 *Number Power,* Bk. 6, pages 7–35, 55–69, 97–106, 125–146 *GED Satellite Program, Mathematics,* pages 21–30, 37–39

Algebra **Pre-Test Score** ☐ **Post-Test Score** ☐

Subskill	TABE, Form 7	TABE, Form 8	Practice and Instruction Pages in This Text	Additional Practice and Instruction Resources
Functions, patterns	4, 5, 37	3, 20	94–95, 98, 104–105*p*	*Critical Thinking with Math,* pages 3–5, 8–12 *Number Power Review,* pages 98–99 *GED Satellite Program, Mathematics,* pages 41–43
Missing element		4	96–97, 104–105*p*	*Number Sense,* Bk. 1, pages 18, 30, 54, 55; Bk. 2, pages 8, 23 *Real Numbers,* Bk. 5, pages 5–10, 12–15
Number sentence	2		99–101*p*, 104–105*p*	*Real Numbers,* Bk. 5, pages 1–2, 11, 16–19, 27–28 *Number Sense,* Bk. 1, pages 10, 25–27, 46–47, 50–56; Bk. 2, pages 1–4, 18–22, 45, 47–48, 50–56 *Number Power,* Bk. 7, pages 70–81
Equations	7	6, 13, 42	99–105*p*	*Real Numbers,* Bk. 5, pages 1–29 *Number Power,* Bk. 7, pages 82–89 *GED Satellite Program, Mathematics,* pages 195–210
Applied algebra	9		100–101*p*, 104–105*p*	*The GED Math Problem Solver,* pages 2–27, 44–53, 59 *Real Numbers,* Bk. 5, pages 17–22, 27–29 *Number Power,* Bk. 7, pages 70–111

Measurement **Pre-Test Score** ☐ **Post-Test Score** ☐

Subskill	**TABE, Form 7**	**TABE, Form 8**	**Practice and Instruction Pages in This Text**	**Additional Practice and Instruction Resources**
Money		2, 47	5, 8–9*p*, 30, 42	*Number Sense,* Bk. 3, pages 3–4, 29–32, 45–54, 56–59; Bk. 4, pages 17–21, 40–60 *Foundations, Mathematics,* pages 74–97 *Breakthroughs in Math / Bk. 1,* pages 14–16, 26–32, 37–43, 50–55, 64–69, 81–83, 91–99, 112–115
Time	24, 27, 29	31	111–112, 115, 119–121*p*	*Real Numbers,* Bk. 4, pages 54–67 *Number Power,* Bk. 9, pages 126–141, 144, 148–151 *Breakthroughs in Math / Bk. 1,* pages 140–153
Length		23	106–115, 120–121*p*	*Real Numbers,* Bk. 4, pages 10–25 *Number Power,* Bk. 9, pages 18–37; Bk. 6, pages 92–94 *Pre-GED Satellite Program Mathematics,* pages 109–116, 120–122, 137–139
Perimeter		22	116, 120–121*p*	*Real Numbers,* Bk. 6, pages 26–32, 41–45, 58–59 *Number Power,* Bk. 4, pages 70–93 *Pre-GED Satellite Program Mathematics,* pages 123–127, 139–142
Area	18		117, 120–121*p*	*Real Numbers,* Bk. 6, pages 33–45, 58–59 *Number Power,* Bk. 4, pages 70–93 *Pre-GED Satellite Program Mathematics,* pages 126–133, 139–142
Volume, capacity	28		106–107, 111–115, 118, 120–121*p*	*Real Numbers,* Bk. 4, pages 26–45 *Number Power,* Bk. 9, pages 88–107; Bk. 6, pages 92–94 *GED Satellite Program Mathematics,* pages 165–176
Surface area		29	117, 120–121*p*	*Pre-GED Satellite Program Mathematics,* pages 126–133, 139–142

Geometry **Pre-Test Score** ☐ **Post-Test Score** ☐

Subskill	TABE, Form 7	TABE, Form 8	Practice and Instruction Pages in This Text	Additional Practice and Instruction Resources
Patterns, shape		50	94	*Critical Thinking with Math,* pages 3–5, 8–12 *Number Power Review,* pages 98–99
Congruency	17		131–132, 135–136*p*	*Real Numbers,* Bk. 6, pages 25, 60–62 *GED Satellite Program, Mathematics,* pages 264–267
Plane figures	45	21	126, 135–136*p*	*Real Numbers,* Bk. 6, pages 1, 16–20 *Number Power,* Bk. 4, pages 82–83 *Pre-GED Satellite Program, Mathematics,* pages 117–120, 122–123
Solid figures	40		133–136*p*	*Number Power,* Bk. 4, pages 120–121 *GED Satellite Program, Mathematics,* pages 225–226 *Pre-GED Satellite Program, Mathematics,* pages 134–136
Angles	19		122–124, 135–136*p*	*Real Numbers,* Bk. 6, pages 3–8, 23 *Number Power,* Bk. 4, pages 6–17, 36–37 *GED Satellite Program, Mathematics,* pages 223, 252–253, 251–260
Triangles	39, 44	18	127–128, 135–136*p*	*Real Numbers,* Bk. 6, pages 21–24 *Number Power,* Bk. 4, pages 34–43 *GED Satellite Program, Mathematics,* pages 257–260
Similarity		34	131–132	*Real Numbers,* Bk. 6, pages 64–68 *GED Satellite Program, Mathematics,* pages 260–264 *The GED Math Problem Solver,* pages 172–173
Parts of a circle		16	129–130, 135–136*p*	*Real Numbers,* Bk. 6, pages 46–49 *Number Power,* Bk. 4, pages 102–105 *The GED Math Problem Solver,* pages 96–101

Computation in Context **Pre-Test Score** ☐ **Post-Test Score** ☐

Subskill	TABE, Form 7	TABE, Form 8	Practice and Instruction Pages in This Text	Additional Practice and Instruction Resources
Whole Numbers	26, 30, 38	24	23–29*p*, 68–70, 81, 92–93*p*, 100–101*p*, 116–118	*Number Power,* Bk. 6, pages 7–35, 55–69, 97–106, 125–131, 135–148 *Breakthroughs in Math / Bk. 1,* pages 28–31, 38–41, 51–55, 65–69, 82–83, 94–97, 113–115, 125–131, 134–139 *The GED Math Problem Solver,* pages 2–17, 50–69, 76–88
Decimals	33	11, 25, 33	34–41*p*, 82, 100–101*p*, 117–118	*Number Sense,* Bk. 3, pages 29–32, 45–59; Bk. 4, pages 18–20, 39–60 *Number Power,* Bk. 6, pages 36–47, 70–72, 107–113, 132–148 *Breakthroughs in Math / Bk. 2,* pages 46–47, 58–63
Fractions	23, 32, 41	46	42, 45, 48, 53–59*p*	*Number Sense,* Bk. 6, pages 25–26, 47–55; Bk. 7, pages 23–25, 45–60 *Number Power,* Bk. 6, pages 48–51, 73–80, 107–113, 132–148 *Breakthroughs in Math / Bk. 2,* pages 80–81, 88–91, 100–103, 107–111, 123
Percent		28, 41, 43	73–79*p*	*Number Power,* Bk. 6, pages 114–124, 132–148 *Breakthroughs in Math / Bk. 2,* pages 123–141 *GED Satellite Program,* pages 144–155

Estimation **Pre-Test Score** ☐ **Post-Test Score** ☐

Subskill	**TABE, Form 7**	**TABE, Form 8**	**Practice and Instruction Pages in This Text**	**Additional Practice and Instruction Resources**
Reasonableness of Answer	20	48	6, 8–9*p*	*Real Numbers,* Bk. 1, pages 1, 2, 4, 16, 23, 30, 45, 46, 50, 52, 59, 68 *Real Numbers,* Bk. 2, pages 1, 17, 23, 33, 37, 39
Rounding		12, 35, 36	7, 8–9*p*, 27, 31, 40–41*p*	*Real Numbers,* Bk. 1, pages 1, 9–11, 17, 24–26, 39–40, 43, 46, 53–56 *Real Numbers,* Bk. 2, pages 10–11, 16, 18, 24–26, 34, 47–49 *The GED Math Problem Solver,* pages 8, 12–13, 56–58, 132–133, 140–141, 186–187
Estimation	12, 16, 21, 22, 42, 46	15, 40, 49	6, 8–9*p*, 27	*Real Numbers,* Bk. 1, pages 1–69; Bk. 2, pages 1–68 *Number Power Review,* pages 8–13, 68–71, 104–107, 146–147 *The GED Math Problem Solver,* pages 8, 12–13, 56–58, 132–133, 140–141, 186–187

The Number System

Place Value

The ten **digits** are 0, 1, 2, 3, 4, 5, 6, 7, 8, and 9. The value of a digit in a number depends on its **place** in that number.

Look at the numbers 17 and 71. They have the same digits, but they are different numbers. That is because the digits are in different places. The number 17 stands for **10 + 1** or **1 ten and 7 ones.** The number 71 stands for **70 + 1** or **7 tens and 1 one.**

The diagram at the right shows the first seven place values for whole numbers. The number 97,403 has digits in the first five places. It has 9 ten thousands, 7 thousands, 4 hundreds, 0 tens, and 3 ones. It has no digit in the millions place or in the hundred thousands place.

millions	hundred thousands	ten thousands	thousands	hundreds	tens	ones
__	__	9	7,	4	0	3

PRACTICE

Fill in the blanks.

1 35,017 has 3 ______ , 5 ______ , 0 ______ , 1 ______ , and 7 ______ .

2 107,403 has ______________________________ .

3 1,500,000 has ______________________________ .

4 Look at the number 3,962,710. What place is the 6 in? ______

5 Look at the number 4,510,329. What place is the 5 in? ______

6 Look at the number 92,613. What digit is in the thousands place? ______

7 Look at the number 3,167,503. What digit is in the hundreds place? ______

8 What is the value of the 5 in 5,432,106? ______

9 What is the value of the 6 in 4,605,912? ______

10 What is the value of the 7 in 117,500? ______

11 What is the value of the 3 in 310,416? ______

12 What is the value of the 5 in 6,750,000? ______

Expanded Form

The expression "90,000 + 7,000 + 400 + 3" is the **expanded form** of the number 97,403. In expanded form, each part of a number has exactly one nonzero digit.

$$
\begin{aligned}
97{,}403 &= (9 \times 10{,}000) + (7 \times 1{,}000) + (4 \times 100) + (0 \times 10) + (3 \times 1) \\
&= 90{,}000 + 7{,}000 + 400 + 0 + 3 \\
&= 90{,}000 + 7{,}000 + 400 + 3
\end{aligned}
$$

There are 0 tens, so the value of the tens place is zero.
The expanded form does not include the tens.

PRACTICE

Write each of these numbers in expanded form. Write the expanded form in two ways, first using the two symbols × and + and then using only the symbol +. For example, you would write 725 as "(7 × 100) + (2 × 10) + (5 × 1)" and as "700 + 20 + 5."

1 5,100 = ______________________________

= ______________________________

2 75,000 = ______________________________

= ______________________________

3 112,025 = ______________________________

= ______________________________

4 43,020 = ______________________________

= ______________________________

5 503,000 = ______________________________

= ______________________________

Naming Large Numbers

When you write a number in words, you use place values.

784	=	700	+	80 + 4
	=	seven hundred		eighty-four

41,032 = 40,000 + 1,000 + 30 + 2
= forty-one thousand, thirty-two

34,502,070 = 30,000,000 + 4,000,000 + 500,000 + 2,000 + 70
= thirty-four million, five hundred two thousand, seventy

The digits in the hundred thousands place, ten thousands place, and thousands place are grouped together. Similarly, the digits in the hundred millions, ten millions, and millions places are grouped together.

Notice that the name "forty-one thousand, thirty-two" does not contain the word "hundreds." That is because there is a zero in the hundreds place. If a number has a zero, you do not say the place value for that digit.

two hundred six = 206

Write "two hundred six" as "206." It is not "26" or "260."

PRACTICE

Write each number below in words.

1 59,505 ______

2 211,525 ______

3 4,500,072 ______

4 901,316 ______

5 23,012 ______

Below, write each number in digits. Watch for number names that skip places. You must put a zero in that position.

6 ninety-one thousand, two hundred one ______

7 sixty-three thousand, four hundred twelve ______

8 three million, four hundred thousand ______

9 eight hundred one thousand, three hundred fifty-six ______

Comparing Whole Numbers

If two whole numbers have different numbers of digits, the number with more digits is greater.

5,013 is greater than 984

If two whole numbers have the same number of digits, start at the left to compare them.

5,862 is greater than 2,974

because 5 is greater than 2.

PRACTICE

Circle the greater number in each problem.

1	12,301	946
2	103,812	946,203
3	55,000	9,415
4	92	112
5	80,912	91,415
6	1,000,000	695,415
7	12,000	9,946
8	45,678	45,078
9	1,512	1,510
10	999,999	100,000

Circle the greatest number in each problem.

11	712	1,042	900
12	888	899	875
13	90	800	812
14	750	92	623
15	511	595	1,596

Arrange each set of numbers from least to greatest.

16 397 115 269 52

17 6,710 952 1,736 9,000

18 17,008 5,107 5,123 952 978

Arrange these digits to make the greatest number possible.

19 2 0 1 9

20 5 8 1 6

21 1 3 7 2

Writing Dollars and Cents

Dollars and cents are written using a **decimal point** (You will learn more about decimals later in this book.). Dollars are shown to the left of the decimal point. Cents are shown to the right of the decimal point. When you say or write an amount in words, you use the word ***and*** to represent the decimal point.

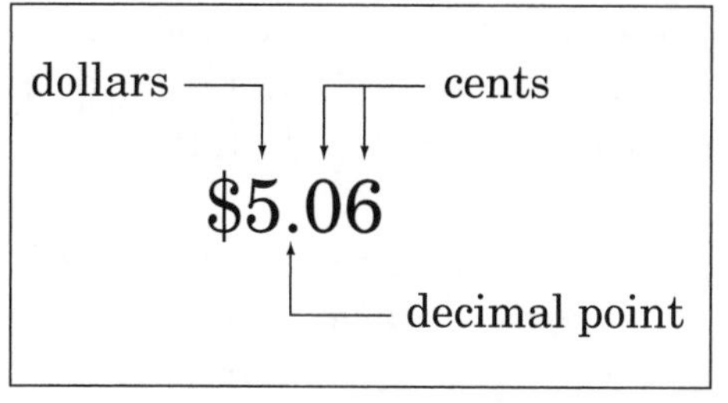

When a dollar amount is written in decimal form, there are always two digits to the right of the decimal point. For example, "two dollars and seven cents" is $2.07. Never write "$2.7" because it is not clear if that means $2.70 or $2.07.

A key to understanding money is to remember that there are *100 cents in a dollar.* Therefore, 50 cents is half of a dollar. To make one dollar it takes four quarters (25 cents each), ten dimes (10 cents each), or 20 nickels (5 cents each).

PRACTICE

1 Fill in the missing parts of this table. Then check your work.

Equal Amounts of Money

cents	dollars	pennies	nickels	dimes	quarters
100	1	100	20	10	4
50					
25				X	
10	$\frac{1}{10}$				X
5	$\frac{1}{20}$			X	X

2 2 quarters and 1 dime is __?__ cents. ______

3 3 dimes and 4 nickels is __?__ cents. ______

4 5 nickels and 1 quarter is __?__ cents. ______

5 1 quarter, 1 nickel, and 1 dime is __?__ cents. ______

Write these amounts in decimal form.

6 one dollar and three cents ______

7 one thousand forty-six dollars and forty-two cents ______

Write each amount below in words.
Sample: $4.03 = four dollars and three cents

8 $5.02 ______________

9 $12.60 ______________

Estimating

Sometimes you need to know *about how much* something costs or *about how much time* something takes. This kind of amount is called an **estimate.** An estimate is a number that is close to an actual or exact amount.

In word problems, the terms *almost, approximately,* and *about* signal that the problem calls for estimation. You should get into the habit of using estimation to check your work after you do any calculation.

PRACTICE

Circle the number for each problem or situation that calls for estimation.

1 The Director of Personnel says that 150 people work in your office. You want to find out for yourself if that number is correct.

2 Your job includes making change.

3 Your child needs 100 milligrams of medicine, but your syringe is in teaspoons. You need to know how many teaspoons are in 100 milligrams.

4 A math problem asks for the approximate number of cans of peaches that are in 12 boxes, where each box contains 52 cans.

5 You want to know if painter A will charge more to paint your house than painter B will charge.

6 A math problem asks about how many eggs make up 9 dozen.

7 Your brother has many books. You are curious to know about how many books he has.

8 You have to balance your checkbook.

You can often use common sense when you estimate. For instance, you know from experience that a hamburger and fries should cost about five dollars, not fifty dollars and not fifty cents. Use common sense estimation to answer each problem below.

9 About how much should 20 gallons of gasoline cost?

F $5.00 **H** $20.00
G $1.50 **J** $60.00

10 About how long would you need for a 500-mile drive on a highway?

A 9 hours **C** 27 hours
B 3 hours **D** 3 days

11 If you earn $15.00 an hour, about how much would you earn in one day?

F $300.00 **H** $100.00
G $25.00 **J** $800.00

12 About how much is the sales tax on a purchase of $15.00?

A $8.00 **C** 3 cents
B $1.25 **D** $10.00

13 About how much would a pound of boiled ham cost?

F 25 cents **H** $3.00
G 15 cents **J** $8.00

Rounding

Another way to estimate the answer to a math problem is to figure it using rounded numbers. A **rounded number** is close to the exact one, but it is easier to work with.

To round a number, think of it as being on part of a hilly number line like this. Numbers ending in 0, 1, 2, 3, and 4 roll back to the nearest low spot. Numbers ending in 5, 6, 7, 8, and 9 roll ahead to the nearest low spot.

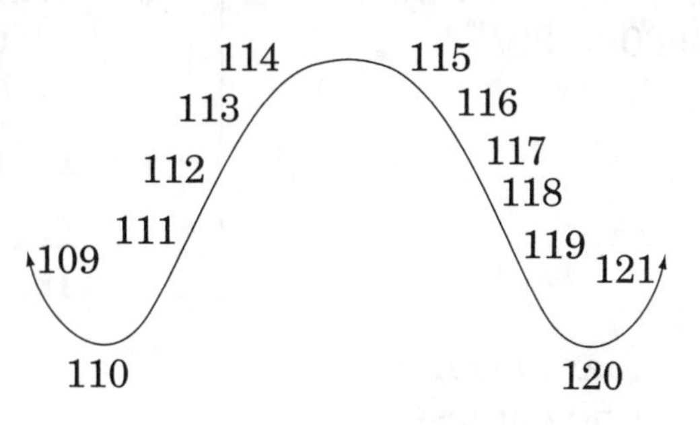

Examples:

111 rounds to 110
118 rounds to 120.
140 rounds to 140.
145 rounds to 150.

PRACTICE

Circle the number that completes each statement.

1	182 rounds to	180	190
2	807 rounds to	800	810
3	99 rounds to	90	100
4	$11.56 rounds to	$11.50	$11.60
5	$9.24 rounds to	$9.20	$9.30
6	1,999 rounds to	2,000	1,990

Round each number to its left-most place.

7 18 rounds to ______

8 32 rounds to ______

9 91 rounds to ______

10 $1.12 rounds to ______

11 86 cents rounds to ______

12 223 rounds to ______

13 698 rounds to ______

You can round to different place values.

- 2,186 rounded to the thousands place is 2,000.
- 2,186 rounded to the hundreds place is 2,200.
- 2,186 rounded to the tens place is 2,190.

To round to a place value, look at the digit just to the right of that place value. If that digit is less than 5, round down. If that digit is 5 or more, round up.

Round each number to the indicated place.

14 3,053 to the hundreds place ______

15 1,906 to the tens place ______

16 25,031 to the thousands place ______

17 967 to the thousands place ______

18 $3.04 to the nearest dollar ______

19 $12.56 to the nearest dime ______

20 $89.00 to the nearest 10 dollars ______

Number System Skills Practice

Circle the letter for the correct answer to each problem.

1 Which of these is another way to write 7,000 + 500 + 30?

A 753
B 7,530
C 7,503
D 70,530

2 Which of these dollar amounts is greater than $1.50 but less than $1.75?

F $1.25
G $1.92
H $1.63
J $1.45

3 Round the number 135,789 to the nearest ten thousand.

A 130,000
B 135,790
C 136,000
D 140,000

4 What is the total value of three quarters, two dimes, and five nickels?

F $0.95
G $0.85
H $1.05
J $1.20

5 Which of these numbers is fifty-four thousand, sixty?

A 54,006
B 54,060
C 5,460
D 5,406

6 Which of these is the correct way to write one hundred twelve dollars and four cents?

F $112 and $\frac{4}{10}$
G $112.4
H $112.04
J $112.40

7 Look at the following four numbers:

25,146 25,832 24,964 25,819

Which choice shows the numbers in order from least to greatest?

A 25,146 25,832 24,964 25,819
B 24,964 25,146 25,832 25,819
C 24,964 25,146 25,819 25,832
D 25,146 25,819 25,832 24,964

8 Which group of coins is worth exactly $\frac{1}{2}$ of a dollar?

F two dimes and a quarter
G a quarter and four nickels
H three nickels and three dimes
J a quarter and five nickels

9 Which of these is the most reasonable estimate of how much it would cost to buy T-shirts for all 20 members of a soccer team?

A $20.00
B $200.00
C $2,000.00
D $20,000.00

Crystal is assistant manager at a dime store. Part of her job is to keep track of the store's sales. This list shows the sales figures for yesterday.

Use the list to answer questions 10 through 14

Yesterday's Sales Figures

Housewares	$9,475.15
Hardware	$12,809.98
Clothing	$10,870.56
Office Supplies	$4,750.42

10 Which department took in the most money yesterday?

F Housewares
G Hardware
H Clothing
J Office supplies

11 Crystal's boss asked her to round the sales figures to the nearest one hundred dollars. Which of these lists shows yesterday's sales rounded to the nearest 100 dollars?

A $9,500, $13,000, $11,000, $5,000
B $9,500, $12,800, $10,800, $4,800
C $9,500, $12,800, $11,000, $4,000
D $9,500, $12,800, $10,900, $4,800

12 Which of these is the most reasonable estimate of the store's total sales yesterday?

F $35,000
G $10,000
H $15,000
J $100,000

13 Crystal calculated that last year's total sales were $19,560,000. What is this amount in words?

A nineteen million, five hundred sixty dollars
B nineteen million, five hundred sixty thousand dollars
C nineteen billion, five hundred sixty million dollars
D nineteen million, five hundred six thousand dollars

14 For which of the following tasks could Crystal most reasonably use an estimate?

F making out a paycheck
G putting prices on blow dryers
H filling out her time card
J reporting how many customers came to the store last year

---

15 Which two numbers will both round to 500?

A 492 and 475
B 515 and 565
C 550 and 450
D 480 and 550

16 Which of these is the most reasonable estimate of how long it would take an adult to walk 5 miles?

F 5 minutes
G 30 minutes
H 2 hours
J 10 hours

Computation

Review of Whole Number Addition

To add large numbers, start by writing them in column form. Start with the two digits in the right column. Add those digits, and write the sum. Then move one column to the left and add the digits in that column. Keep this up until you have added all the columns.

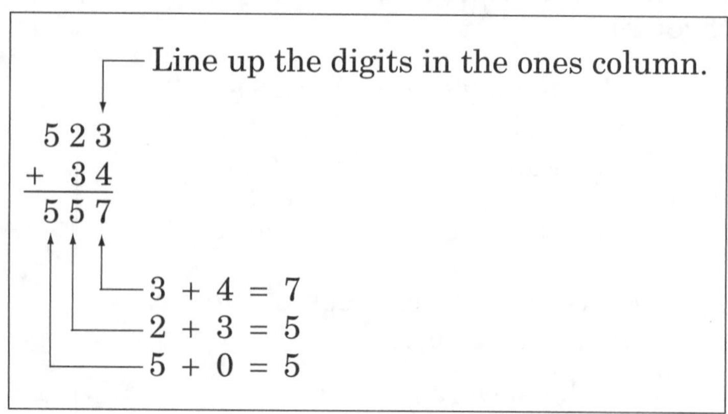

$$\begin{array}{r} \$1.14 \\ +\ 3.24 \\ \hline \$4.38 \end{array}$$

If the numbers in a problem have decimal points or labels, repeat them in your answer. In an addition problem, the decimal points line up vertically.

To add several numbers, add two digits at a time. Keep adding the "next" digit until you complete the column. Then go on the the next column of digits.

$$\begin{array}{r} 21 \\ 23 \\ 2 \\ +\ 11 \\ \hline 57 \end{array}$$

Add the first two digits: $1 + 3 = 4$

Think: $4 + 2 = 6$

$6 + 1 = 7$

Write the 7. Then think: $2 + 2 = 4$, $4 + 1 = 5$

Write the 5.

PRACTICE

Solve each problem below.

1 81 + 5 = ______

2 $12.15 + $6.52 = ______

3 517 + 40 = ______

4 19 min + 110 min = ______

5 621 gal + 52 gal = ______

6 12,402 + 491 = ______

7 80,215 + 4,552 = ______

8 711 + 350 = ______

Check your work on this page using subtraction. For example, subtract 5 from your answer to problem 1. The result should be 81.

9 23 + 13 + 150 = ______

10 701 + 50 + 110 = ______

Carrying

When the sum of a column is more than 9, write the ones-place digit in the answer. Then "carry" the tens-place digit to the top of the tens column. Repeat this process each time the sum of a column is greater than 9.

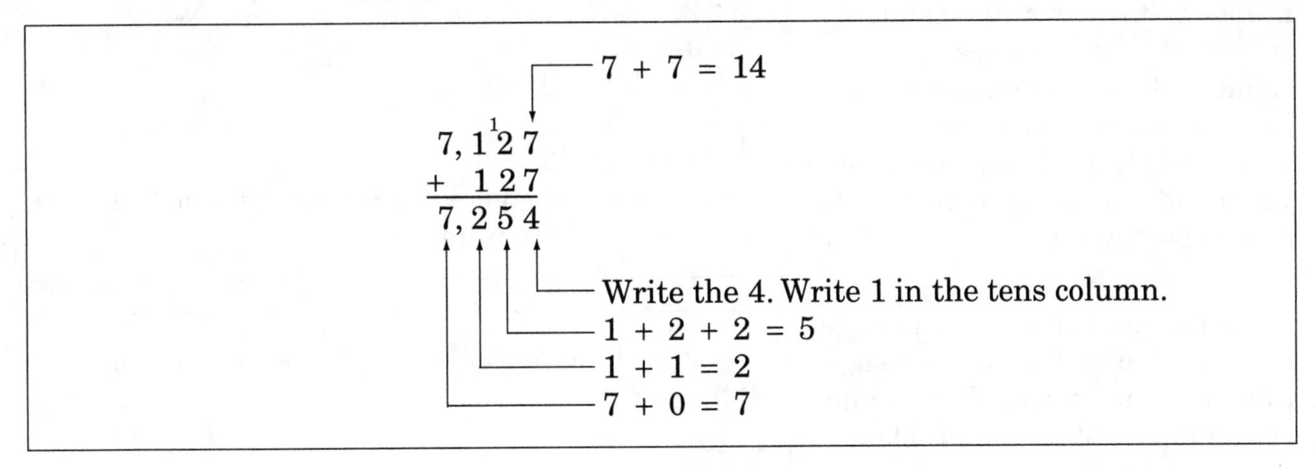

PRACTICE

Rewrite each problem in column form, and find the sum. Check your work by rounding the numbers and adding.

1 $16.00 + $19.00 = ______

2 1,076 + 62 = ______

3 $59.00 + $85.00 = ______

4 $7.50 + $11.75 = ______

5 5,170 + 490 = ______

6 497 + 612 = ______

7 809 + 69 = ______

8 97 + 156 = ______

9 7,175 + 233 = ______

10 675 + 2,050 = ______

11 712 + 199 = ______

12 69 + 1,678 = ______

Review of Whole Number Subtraction

To subtract, write the numbers in column form. The number you are subtracting *from* must be the top number. Starting at the right, subtract the digits in each column. If the difference in a column is zero, be sure to write a zero below the column. However, you *do not* write a zero at the left of a whole number.

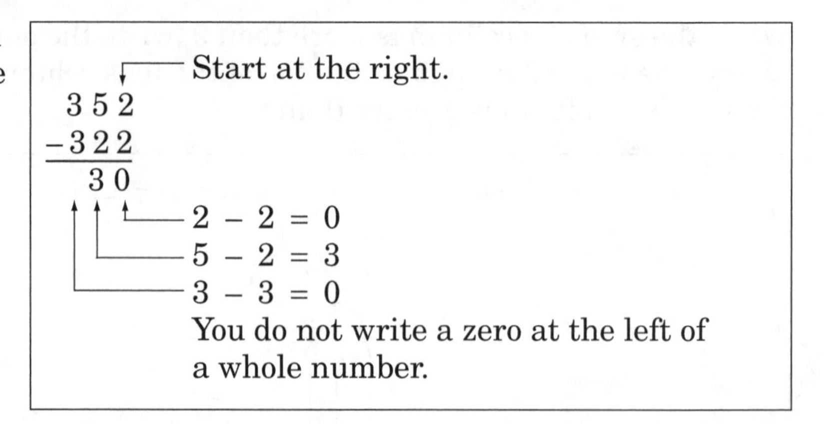

If the numbers in a problem have labels or decimal points, repeat them in your answer. The decimal points in a subtraction problem line up vertically.

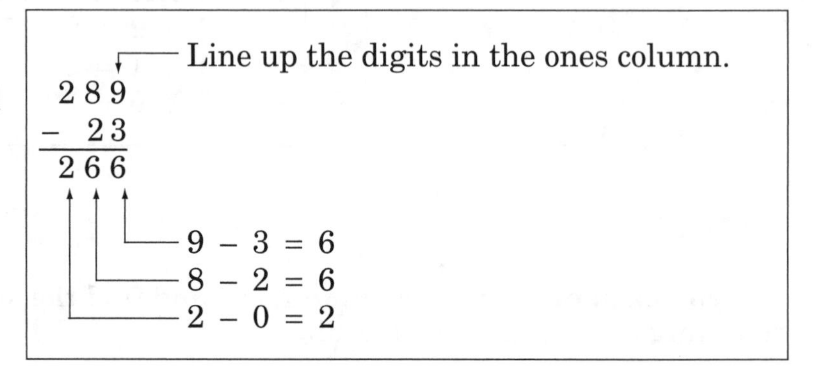

PRACTICE

Rewrite each problem in column form. Solve the problem. Then use addition to check your answer. The sum of the two bottom numbers should be the same as the top number.

1 311 miles − 110 miles = ______

2 555 − 224 = ______

3 $95.00 − $73.00 = ______

4 76 in. − 23 in. = ______

5 $157.00 − $25.00 = ______

6 1,670 − 420 = ______

7 892 − 131 = ______

8 $42.49 − $30.15 = ______

9 75 − 24 = ______

10 422 − 310 = ______

Borrowing

In each problem below, sometimes the digit in the bottom number is greater than the digit above it. Each time you "borrow," you replace the number you are *borrowing from* with a digit that is one less than that digit. Then you replace the digit that is too small with a number that is ten more. Be sure to write neatly.

```
      12
    7 2̸ 11
 1, 8̸ 3̸ 1̸
 –   2 4 9
 1,  5 8 2
```

- Think of 30 + 1 as 20 + 11. Then 11 – 9 = 2.
- Think of 800 + 20 as 700 + 120. Then 12 – 4 = 8.
- 7 – 2 = 5
- 1 – 0 = 1

The following example illustrates how to borrow from a digit that is zero.

```
   9
 3 1̸0̸ 17
 4̸ 0̸ 7̸
 –    9
 3 9 8
```

You can borrow from zero.

400 + 0 + 7 = 300 + 100 + 7
= 300 + 90 + 17

- 17 – 9 = 8
- 9 – 0 = 9
- 3 – 0 = 3

Here is a shortcut when borrowing from a row of zeros. Write a digit one less than the first nonzero digit. Then write 10 over the digit you need and write 9 over each digit in between.

```
 3 9  9 9 10
 4̸ 0̸, 0̸ 0̸ 0̸
 –      5 2 8
 3 9, 4 7 2
```

PRACTICE

Solve each subtraction problem. Use rounded numbers to check your work.

1 882 – 29

2 281 – 85

3 500 – 76

4 \$5.21 – 1.32

5 3,100 – 5

6 2,500 – 250

7 \$32.10 – 5.90

8 \$100.00 – 51.00

9 918 – 99

10 \$40.00 – \$2.50 = ______

11 596 – 59 = ______

12 8,000 – 12 = ______

Multiplying by a One-Digit Number

Here is a multiplication problem in column form. Multiply the bottom number times each digit in the top number, working from right to left.

$$\begin{array}{r} 134 \\ \times \quad 2 \\ \hline 268 \end{array}$$

← Multiply by this digit.

2 × 4 = 8
2 × 3 = 6
2 × 1 = 2

If one of the numbers in a problem has a label, repeat it in your answer. Also, remember that dollar amounts always have two digits to the right of the decimal point.

PRACTICE

Rewrite each multiplication problem in column form. Then solve the problem. Review your work to make sure it is correct.

1 124 × 2 = ______

2 3 × 310 = ______

3 5 × 511 = ______

4 81 × 6 = ______

5 202 × 4 = ______

6 $4.21 × 2 = ______

7 $7.04 × 2 = ______

8 6,102 × 4 = ______

9 $21.00 × 8 = ______

10 3 × 523 = ______

11 5 × 60 = ______

12 310 × 2 = ______

13 511 × 3 = ______

14 132 × 2 = ______

15 213 × 3 = ______

Multiplying by a Two-Digit Number

In the problem below, you are multiplying by the two-digit number 23.

- First, multiply the top number by the ones digit of the bottom number.
- Next, multiply the top number by the tens digit in the bottom number. Start your answer in the tens column, leaving a blank space in the ones place.
- Finish the problem by adding the two products.

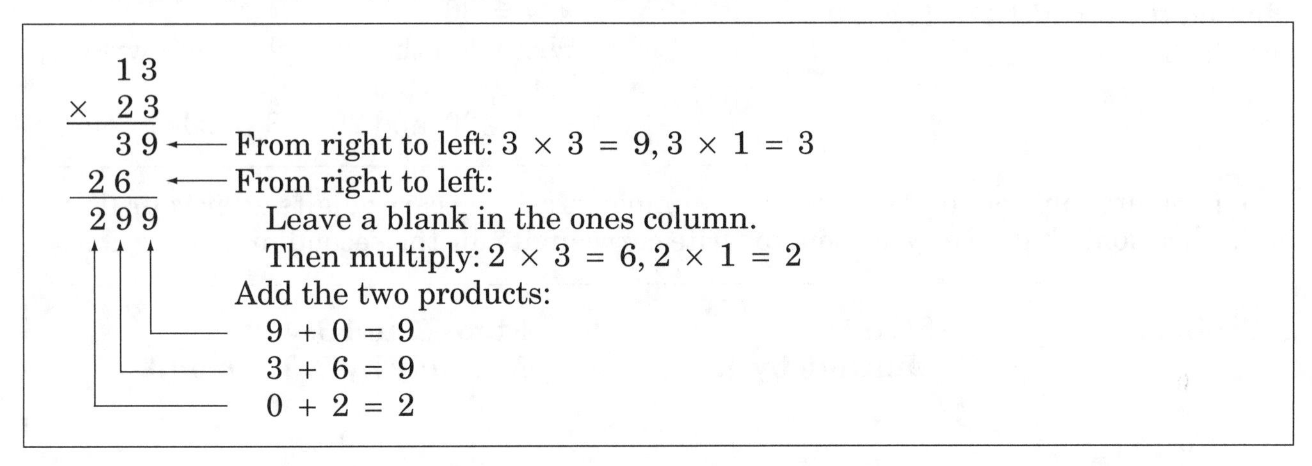

When you multiply a dollar amount times a whole number, ignore the decimal point at first. After you find the product, write a decimal point two places from the right of your answer. *Remember:* in a dollar amount, there are always two digits to the right of the decimal point.

PRACTICE

Rewrite each multiplication problem in column form. Then solve the problem. Use rounding to check your work.

1 $312 \times 30 =$ ______

2 $51 \times 22 =$ ______

3 $70 \times 21 =$ ______

4 $3{,}021 \times 30 =$ ______

5 $3{,}130 \times 12 =$ ______

6 $\$2.10 \times 11 =$ ______

7 $202 \times 13 =$ ______

8 $\$3.30 \times 23 =$ ______

9 $20 \times 15 =$ ______

10 $100 \times 21 =$ ______

11 $\$1.30 \times 23 =$ ______

12 $\$1.60 \times 11 =$ ______

Carrying When You Multiply

In this multiplication problem, the product in the ones column is a two-digit number. You write the digit 6 in the ones place as part of your answer. Then you write the digit 3 above the tens place. After you find the product 4 × 5, you add the 3.

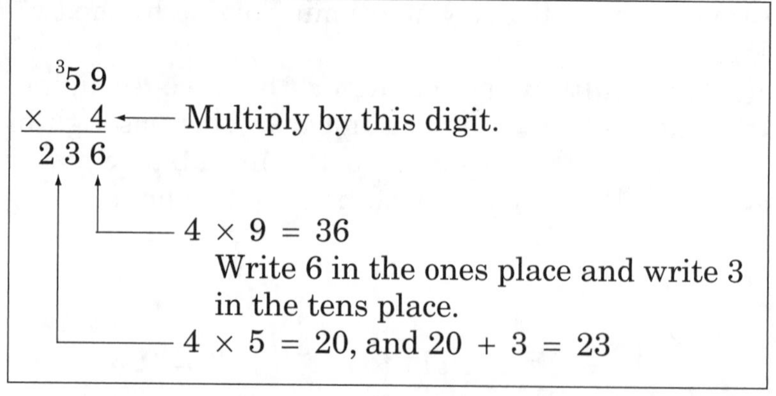

When you are multiplying by a two-digit number, erase the carry digits after your first multiplication. That gives you room to write carry digits for the second multiplication.

Problem:	**Step 1:** **Multiply by 4.**	**Steps 2 and 3:** **Multiply by 2. Then add.**
35 × 24	²35 × 24 140	¹35 × 24 140 70 840

PRACTICE

Solve each problem. Then check your work. *Remember:* Multiply *before* you add the digit being carried.

1 25×5

2 75×14

3 512×37

4 69×3

5 $\$1.72 \times 51$

6 $3{,}045 \times 13$

7 $1.48 × 3 = ______

8 42 × 8 = ______

9 135 × 90 = ______

10 $309 \times 17 =$ ______

11 $\$1.53 \times 15 =$ ______

12
$$\begin{array}{r} \$3.40 \\ \times\ 12 \\ \hline \end{array}$$

13
$$\begin{array}{r} 592 \\ \times\ 24 \\ \hline \end{array}$$

14
$$\begin{array}{r} 315 \\ \times\ 25 \\ \hline \end{array}$$

15
$$\begin{array}{r} \$1.90 \\ \times\ 32 \\ \hline \end{array}$$

16 $198 \times 60 =$ ______

17 $\$2.25 \times 3 =$ ______

18 $816 \times 7 =$ ______

19 $391 \times 12 =$ ______

20 $825 \times 50 =$ ______

21 $\$5.35 \times 36 =$ ______

22 $326 \times 45 =$ ______

23 $\$7.17 \times 34 =$ ______

24 $58 \times 52 =$ ______

25 $87 \times 22 =$ ______

26 $3{,}311 \times 15 =$ ______

27 $\$20.50 \times 17 =$ ______

28 $2{,}018 \times 45 =$ ______

29 $1{,}091 \times 63 =$ ______

30 $\$20.13 \times 15 =$ ______

Using the Division Bracket $\overline{)}$

To write a division problem using the bracket $\overline{)}$, the number you are *dividing up* (the **dividend**) goes inside the bracket. The number you are *dividing by* (the **divisor**) goes to the left of the bracket. The answer in a division problem is called the **quotient.**

Starting at the left, divide the divisor 2 into each digit of the dividend 480.

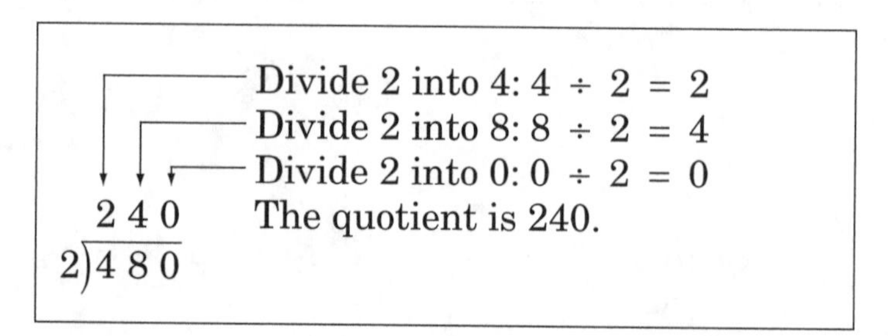

In the problem $6\overline{)426}$, 6 is greater than 4. To start, divide 6 into the first two digits of 426.

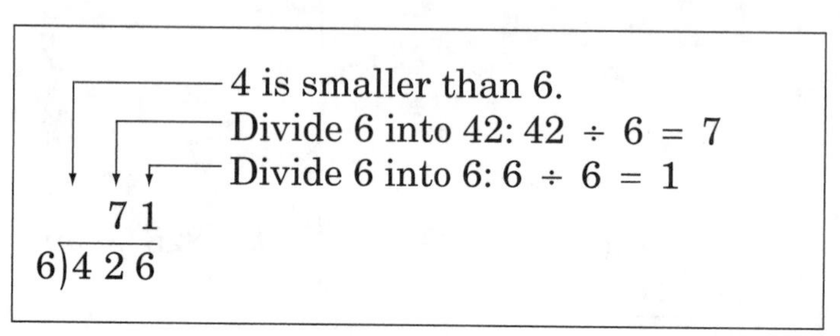

After you write the first digit in the quotient, use a zero as a placeholder whenever a digit in the dividend (inside the bracket) is less than the divisor.

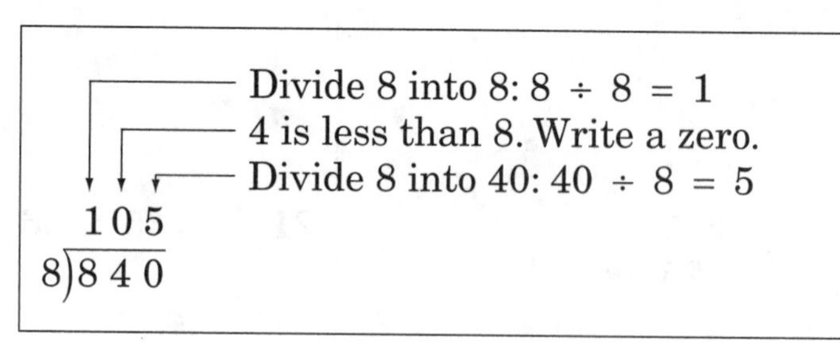

PRACTICE

Solve each problem. Divide into every digit under the division bracket, even if that digit is a zero. To check your answers, multiply the quotient times the divisor. The product should be the dividend of the original problem.

1 $2\overline{)248}$

2 $8\overline{)248}$

3 $3\overline{)369}$

4 $3\overline{)126}$

5 $3\overline{)324}$

6 What is 48 divided by 4?

7 What is 628 divided by 2?

8 What is 189 divided by 9?

9 What is 624 divided by 6?

Dividing with a Remainder

One number cannot always be evenly divided by another. For instance, 19 objects cannot be divided into 5 equal piles. We say that such a problem has a **remainder.** If you divide 19 objects into 5 piles of 3, there will be 4 objects left over. We say that "19 divided by 5 is 3, with a remainder of 4."

$19 \div 5 = 3 \text{ r } 4$

To solve a problem with a remainder, you must find the largest whole number that goes into the dividend. The next step is to multiply that digit by the divisor. The last step is to subtract.

$$\begin{array}{r} 58 \text{ r } 3 \\ 4\overline{)235} \\ \underline{20} \\ 35 \\ \underline{32} \\ 3 \end{array}$$

1. Divide 4 into 23. Write 5 in the quotient.
2. Multiply: $5 \times 4 = 20$. Write 20.
3. Subtract: $23 - 20 = 3$. Bring down the next digit 5.
4. Divide 4 into 35. Write 8 in the quotient and then multiply: $8 \times 4 = 32$.
5. Subtract: $35 - 32 = 3$.

The quotient is 58 r 3.

PRACTICE

Solve each problem. (*Hint:* The correct remainder is always less than the divisor.) Check your work by multiplying the quotient times the divisor. Then add the remainder. The result should be the dividend from the original problem.

1 $2\overline{)27}$

2 $3\overline{)32}$

3 $7\overline{)79}$

4 $6\overline{)13}$

5 $8\overline{)85}$

6 What is 18 divided by 4?

7 What is 20 divided by 9?

8 What is 190 divided by 9?

9 What is 614 divided by 6?

10 What is 725 divided by 7?

11 $8\overline{)1629}$

12 $5\overline{)1523}$

13 What is 985 divided by 9?

14 What is 629 divided by 6?

Writing the Steps of a Division Problem

In the problem below, the divisor 6 does not evenly divide the first two digits of the dividend 474.

Here is a step-by-step method for solving such a problem.

$$\begin{array}{r} 79 \\ 6\overline{)474} \\ \underline{42} \\ 54 \\ \underline{54} \\ 0 \end{array}$$

1. Divide 6 into 47. Write 7 in the quotient.
2. Multiply: $7 \times 6 = 42$. Write 42.
3. Subtract: $47 - 42 = 5$. Bring down the digit 4.
4. Divide: $54 \div 6 = 9$. Write 9 in the quotient.
5. Multiply: $9 \times 6 = 54$. Subtract; $54 - 54 = 0$.

The quotient is 79, with no remainder.

PRACTICE

Solve each problem. Check your work by multiplying the quotient times the divisor.

1 $2\overline{)172}$

2 $4\overline{)224}$

3 $5\overline{)805}$

4 $6\overline{)324}$

5 $7\overline{)91}$

6 What is 896 divided by 8?

7 What is 231 divided by 3?

8 What is 204 divided by 6?

9 What is 1,326 divided by 6?

10 What is 161 divided by 7?

11 $5\overline{)2{,}750}$

12 $4\overline{)152}$

13 What is 198 divided by 9?

14 What is 2,706 divided by 6?

15 What is 917 divided by 7?

16 What is 140 divided by 4?

Long Division

A division problem can become complicated. You should write each step neatly.

Look at the problem below. Be sure to notice that after the first three steps, you always follow the same four-step pattern:

- Bring down the next digit.
- Divide.
- Multiply.
- Subtract.

This sequence of steps is called **long division.**

$$\begin{array}{r} 345 \\ 4\overline{)1380} \\ \underline{12} \\ 18 \\ \underline{16} \\ 20 \\ \underline{20} \\ 0 \end{array}$$

1 Divide 4 into 13. Write 3 in the quotient. **2** Multiply: $3 \times 4 = 12$. **3** Subtract: $13 - 12 = 1$.	**4** Bring down the next digit 8. **5** Divide 4 into 18. Write 4 in the quotient. **6** Multiply: $4 \times 4 = 16$. **7** Subtract: $18 - 16 = 2$.	**8** Bring down the next digit 0. **9** Divide 4 into 20. Write 5 in the quotient. **10** Multiply: $5 \times 4 = 20$. **11** Subtract: $20 - 20 = 0$.

PRACTICE

Use long division to solve these problems. Show all your work, and check your answers using multiplication.

1 $2\overline{)332}$

2 $5\overline{)720}$

3 $7\overline{)931}$

4 $6\overline{)2{,}652}$

5 $7\overline{)1{,}505}$

6 $3\overline{)9{,}144}$

Dividing by a Two-Digit Number

When you divide by a two-digit number, the first step is to find the digit for the quotient. Look at these two examples.

Example 1

$$\begin{array}{r} 21 \\ 15\overline{)315} \\ \underline{30} \\ 15 \\ \underline{15} \\ 0 \end{array}$$

1 Divide 31 by 15. Write 2.
2 Multiply: 2 × 15 = 30.
3 Subtract: 31 − 30 = 1.
4 Bring down the next digit 5.
5 Divide 15 by 15. Write 1.
6 Multiply: 1 × 15 = 15.
7 Subtract: 15 − 15 = 0.

Example 2

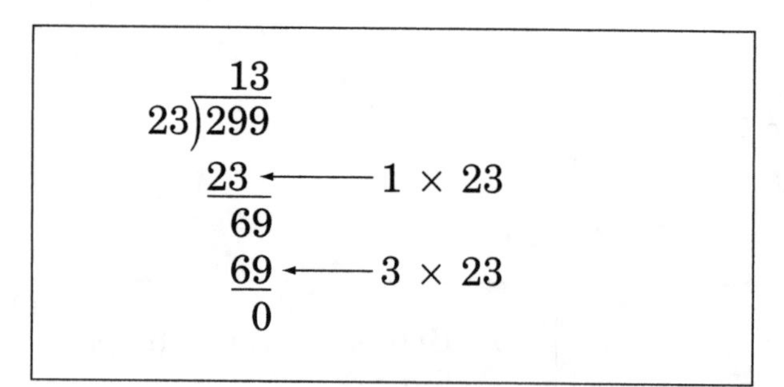

Important: The results of the "multiply" step *must not* be greater than the number above it. If it *is* greater, do the step again using a lesser digit in the divisor.

PRACTICE

Solve each problem below. Then check your work. *Hint:* These problems do not have remainders.

1 $13\overline{)260}$

2 $50\overline{)600}$

3 $37\overline{)740}$

4 $15\overline{)480}$

5 $20\overline{)460}$

6 $24\overline{)720}$

7 $43\overline{)172}$

8 $16\overline{)320}$

9 $25\overline{)525}$

Solving Word Problems

An important part of solving any word problem is to decide whether you should add, multiply, subtract, or divide. Look for these clues.

- If you put different amounts together, you will add.
- If you compare one amount to another amount, you will subtract or divide.
- If you take something away from something else, you will subtract.
- If there are several of something, you will multiply or divide.

Here are some "signal words" for the four math operations of addition, subtraction, multiplication, and division.

Addition	Subtraction	Multiplication	Division
plus sum total added to altogether in all combined increased by	minus difference take away subtract left over how much change more than less than decreased by	times multiplied by product twice, three times, and so on apiece each	divide apiece each per

PRACTICE

For each problem, tell whether you will add, subtract, multiply, or divide. You do not have to solve the problem.

1 You buy a book for $8.95. You pay with a 10-dollar bill. What should you do to find how much change should you get back?

A Add. **C** Multiply.
B Subtract. **D** Divide.

2 It will cost $350.00 for the ladies club to go to a dinner theater. There are 12 ladies in the club. How much will it cost per lady?

F Add. **H** Multiply.
G Subtract. **J** Divide.

3 You buy a new dryer for $250.00. Tax is $22.47. How much do you owe in all?

A Add. **C** Multiply.
B Subtract. **D** Divide.

4 Your gas tank holds 20 gallons. You fill it up and drive 420 miles before the gas gauge reads empty. How many miles did you drive per gallon of gas?

F Add. **H** Multiply.
G Subtract. **J** Divide.

5 Museum tickets are $8.00 per person. How much would it cost to take a family of five?

A Add. **C** Multiply.
B Subtract. **D** Divide.

6 You have a board that is 16 inches long. You cut off a piece that is 5 inches long. How much of the board is left?

F Add. **H** Multiply.
G Subtract. **J** Divide.

Another important part of solving a word problem is to find the specific information that you need.

Always read and reread the problem carefully, and ask yourself, "What is the particular question in the problem?" Use all the clues in the problem. For example, if an addition problem asks for a dollar amount, you know that you will be adding amounts of money.

PRACTICE

Each word problem below contains more information than you need. Circle the numbers you need to solve each problem. Then circle the letter for the operation you would use to solve the problem.

7 Becky agreed to make 12 dozen cookies for a bake sale, but her sister ended up making half of them. It took Becky 10 minutes to mix one batch of cookies. How long will it take her to mix 6 batches?

A Add. **C** Multiply.
B Subtract. **D** Divide.

8 Of the 50,000 people who live in Bakersfield, 5,156 of them work for the Ford plant. Also, 15,125 people moved to Bakersfield within the last 3 years. How many of the people living in Bakersfield have been there for more than 3 years?

F Add. **H** Multiply.
G Subtract. **J** Divide.

9 Rachel wants to take her scout troop to Washington, D.C. Airfare would be $125 per scout. The hotel would cost $60.00 per person per night. How much would it cost each scout to stay at the hotel for 3 nights?

A Add. **C** Multiply.
B Subtract. **D** Divide.

10 Ron is 65 years old. His wife was born 3 years after he was born. When she was 25, she had their first child. How old was she on that child's 15th birthday?

F Add. **H** Multiply.
G Subtract. **J** Divide.

Each word problem below does not give enough information. On the blank lines, tell what else you would need to know before you could solve each problem.

11 Phu rented a Halloween costume for 3 days. How much did he owe?

You also need to know

12 The company bought lunch for the secretaries in Jan's office. Each lunch cost $18.00. How much did the company spend?

You also need to know

13 Chris likes a carpet that costs $19.00 a square yard. How much would it cost to buy this carpeting for his living room?

You also need to know

Set-up Problems

To **set up a word problem,** or to **write a set-up** for a problem, means to write an addition, subtraction, multiplication, or division problem. For subtraction and division problems, the order of the numbers is important.

PRACTICE

Set up each problem. Then solve the problem.

1 A plumber spent 3 hours at Wayne's house. He charged $120 for labor. How much did the plumber charge per hour?

2 The gas tax in Rhode Island is 23 cents per gallon. If you buy 20 gallons of gas, how much tax will you pay?

3 Lance loaded 89 boxes into his truck in the morning. During the day, he unloaded 53 boxes. How many boxes were left?

4 Marcus's company was paid $3,500 for replacing a sewer line. Marcus paid his workers $1,800. He spent $700 on materials. How much money did Marcus make?

Sergio's regular pay rate is 18 dollars an hour. When he works overtime, his pay rate is 9 dollars more than his regular rate. Last week he worked 8 hours every weekday, plus 2 hours overtime on Friday.

Use this information to set up and solve problems 5 through 8.

5 How much does Sergio earn per hour when he works overtime?

6 How many hours did Sergio work last week? *(This problem needs two steps to be solved.)*

7 How much does Sergio earn in an 8-hour day with no overtime?

8 If Sergio gets a raise of $2.00 per hour, how much will he make in a regular 40-hour week? *(This problem needs two steps to be solved.)*

Solving Two-Step Word Problems

Some problems need two steps to be solved. As one example, you might have to add two numbers, then multiply the sum by a number. As another example, you might have to multiply and then divide. Any combination of operations is possible. There is no special trick to solving two-step problems. Just read them carefully, looking for clues and signal words.

PRACTICE

Each problem below needs two steps to be solved. Circle the letter for the choice that describes how to solve the problem.

1 Mrs. Carr's class is taking a field trip, and five parents have offered to help drive. Four of the parents have vans that hold nine children each. The other parent has a car that can only carry four children and the driver. Together, how many children can these parents drive?

A Add 9 and 4. Then multiply by 5.
B Multiply 4 by 9. Then add 4.
C Add 4 and 9. Then add 4.

2 Terry has $6.84 in his wallet. He gets $40.00 from an ATM machine. Then he spends $15.95. How much money does he have left?

F Add $6.84 and $40.00. Then subtract $15.95.
G Subtract $6.84 from $40.00. Then subtract $15.95.
H Add $6.84 and $15.95. Then subtract the sum from $40.00.

3 Wendell buys 3 pounds of hamburger for $2.56 a pound, plus a box of cereal for $3.95. How much did he spend? *(Ignore the tax.)*

A Add $3.95 and $2.56. Then multiply by 3.
B Add 3 and $2.56. Then add $3.95.
C Multiply $2.56 by 3. Then add $3.95.

Solve each two-step problem below.

4 Ellen knows that it will take her 15 minutes to mix a batch of cookies, and then 10 minutes per pan to bake them. If a batch of cookies makes 3 pans, how long will it take Ellen to mix and bake them all?

5 Eric buys 3 pounds of chocolates for $4.50 a pound. The tax on his purchase is $1.25. How much does he spend?

6 Tina had $2,546 in a savings account. She deposited $520 dollars, then took out $40.00. What was the balance in her savings account?

7 Jorge has 28 miniature cars and 16 stuffed animals that he has saved from his childhood. He divides the toys evenly among his 4 children. How many toys does each child get?

Using Estimation in Word Problems

The words *about, approximately,* and *almost* usually signal that a word problem calls for estimation. When a problem tells you to estimate, set up the problem as you would any other problem. Then round the numbers before you do any calculating.

PRACTICE

Circle the letter for the number sentence that shows the *best* way to solve each problem. *Hint:* Some problems call for estimation, and some do not.

1 In an election, 7,845 people voted for Miller and 3,182 voted for her opponent. About how many people voted in that race?

A 7,845 + 3,182 = _____
B 7,000 + 3,000 = _____
C 8,000 + 3,000 = _____

2 A sweater costs $49.99. Tax is $0.12 per dollar. How much would it cost to buy this sweater?

F $49.99 × $0.12 = _____
G $49.99 + ($49.99 × $0.12) = _____
H $49.99 + $0.12 = _____

3 Glenn makes $26,412 a year, plus about $3,000 a year in bonuses. About how much does he make per month?

A ($26,412 + $3,000) ÷ 12 = _____
B ($26,000 + $3,000) ÷ 12 = _____
C ($27,000 + $3,000) ÷ 12 = _____

4 Naomi spent 12 minutes on a stair machine, 9 minutes on a bike, and 30 minutes in aerobics class. About how long did she exercise?

F 15 + 10 + 30 = _____
G 15 + 10 + 50 = _____
H 10 + 10 + 30 = _____

Use estimation to solve each of the following problems.

5 It took Linda 3 hours to drive 178 miles. Approximately how many miles did she drive per hour? *(Round 178 to the tens place.)*

6 This month, Carlyle spent $72.50 to get his car tuned up, $125.12 to get new brake pads, and $119.95 to get a new alternator. To the nearest dollar, how much did he spend?

7 Arlene took 6 dresses to the dry cleaners. They charge $2.95 per dress. To the nearest dollar, how much will Arlene be charged?

8 George has a library book that is 23 days overdue. Library fines are 25 cents a day. About how much will George owe? *(Round each number to the tens place.)*

Computation Skills Practice

Circle the letter for the best answer to each problem. Try crossing out unreasonable answers before you start to work on each problem.

1

$$\begin{array}{r} 315 \\ \times\ 9 \\ \hline \end{array}$$

A 2,715
B 2,735
C 2,845
D 2,835
E None of these

2

$272 \div 8 =$ _____

F 34
G 30 r 2
H 44
J 32
K None of these

3

$231 \times 50 =$ _____

A 1,155
B 11,550
C 10,550
D 1,055
E None of these

4

$$\begin{array}{r} 3801 \\ \times\ 13 \\ \hline \end{array}$$

F 39,253
G 15,204
H 47,413
J 49,413
K None of these

5

$14\overline{)420}$

A 3
B 33
C 29
D 40
E None of these

6

$3\overline{)672}$

F 224
G 220 r 2
H 230
J 223
K None of these

Study this mileage chart. Then do Numbers 7 through 9.

Kirkwood to St. Louis	12 miles
St. Louis to Ft. Leonardwood	128 miles
St. Louis to Rolla	106 miles
St. Louis to Dixon	128 miles
Dixon to Ft. Leonardwood	20 miles
Dixon to St. Robert	17 miles
St. Robert to Ft. Leonardwood	5 miles

7 How many miles is the trip from St. Louis to Ft. Leonardwood if you stop off in Dixon on the way?

A 128 miles
B 148 miles
C 108 miles
D Not enough information is given.

8 Kirkwood lies between St. Louis and Rolla on Highway 44. How far is Kirkwood from Rolla?

F 118 miles
G 94 miles
H 128 miles
J This cannot be determined.

9 You are traveling to Ft. Leonardwood from Dixon. How many miles would you add to your trip if you stopped off in St. Robert along the way?

A 0 miles
B 3 miles
C 2 miles
D 5 miles

10

35 × 8 = _____

F 280
G 240
H 28
J 285
K None of these

11

842 ÷ 6 = _____

A 14 r 2
B 107
C 140 r 2
D 130 r 2
E None of these

12

$3\overline{)105}$

F 35
G 30 r 5
H 3 r 5
J 33
K None of these

13

$$\begin{array}{r} \$8.95 \\ +\ \$5.25 \\ \hline \end{array}$$

A $14.10
B $14.15
C $13.10
D $13.20
E None of these

14

435 × 60 = _____

F 2,610
G 25,800
H 42,600
J 26,100
K None of these

15

84 ÷ 12 = _____

A 70
B 7
C 60
D 7 r 2
E None of these

16

$$\begin{array}{r} 53 \\ \times\ 21 \\ \hline \end{array}$$

F 1,013
G 159
H 1,113
J 213
K None of these

17 Lane must buy three hot dogs that cost $1.99 each. Which of these number sentences could he use to find the total cost?

A $1.99 + 3 = _____
B $1.99 × 3 = _____
C $1.99 ÷ 3 = _____
D $1.99 – 3 = _____

18 Luke and Garrett earn $500.00 moving a family across town. They use $40.00 of that money to pay for renting the moving van. If they split the rest of the money evenly, how much money will each man get?

F $250.00
G $210.00
H $460.00
J $230.00

19 Irena has sold 25 makeup kits, and she gets $10.00 for each one. She spent $100.00 to buy the samples that she shows to her customers. What must Irena do to figure out how much money she has made in her business?

A Multiply $100.00 by 25. Then subtract $10.00 from the product.
B Divide $100.00 by $10.00. Then multiply the quotient by 25.
C Multiply $10.00 by 25. Then subtract $100.00 from the product.
D Multiply $10.00 by 25. Then add $100.00 to the product.

Decimals

Decimal Place Values

There are **place values** to the right of the ones place. In a number such as 2.63, the period is called a **decimal point** and the first four places to the right of the decimal point are **tenths, hundredths, thousandths,** and **ten-thousandths.** The number 2.63 is called a decimal fraction or, more simply, a **decimal.**

ones		tenths	hundredths	thousandths	ten-thousandths
2	.	6	3	__	__

Two and sixty-three hundredths

In the decimal 2.63, the digit 6 represents six tenths and the digit 3 represents three hundredths.

When you say or write a decimal in words, you use the word **and** to represent the decimal point. So the decimal 2.63 is *"two* **and** *sixty-three hundredths."*

Amounts of money use decimals. A cent is a special name for *one hundredth of a dollar.* So another way to say $9.42 is "nine whole dollars and 42 hundredths of a dollar."

> The 4 in $9.42 refers to tenths of a dollar.
>
> The value of the 4 is $\frac{4}{10}$.
>
> The 2 in $9.42 refers to hundredths of a dollar.
>
> The value of the 2 is $\frac{2}{100}$.

PRACTICE

Fill in the blanks below.

1 In the number 1.93, what place is the 3 in? ______

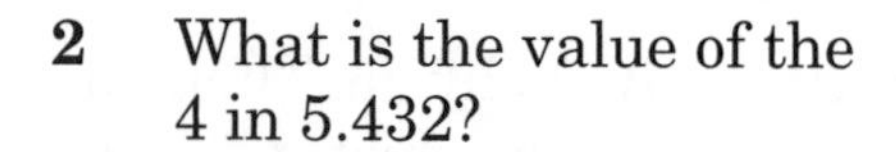

2 What is the value of the 4 in 5.432? ______

3 What is the value of the 9 in 0.09? ______

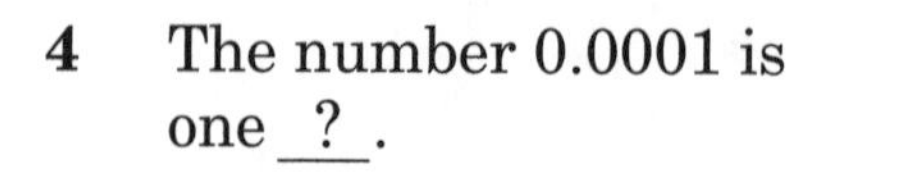

4 The number 0.0001 is one _?_. ______

When you say a decimal name aloud, the place value you say is the place value of the digit *farthest to the right.*

For example, for the decimal 0.083, the digit 3 is the farthest digit to the right. It is in the thousandths place. So 0.083 is "eighty-three thousandths."

5 0.045 is forty-five _?_.

____

6 0.22 is twenty-two _?_.

____

Comparing Decimal Numbers

To compare two decimal numbers, start by lining up the decimal points. Then, moving from left to right, compare the digits.

0.027	2.105	15.063
↕	↕	↕
0.008	12.182	15.017

2 is greater than 0.
The top number is greater.

1 is greater than 0.
The bottom number is greater.

6 is greater than 1.
The top number is greater.

As another example, compare 0.1 and 0.008. You can think of 0.1 as 0.100. In the tenths place, 1 is greater than 0. That means 0.1 is greater than 0.008. With decimals, any nonzero digit *close to* a decimal point represents a greater value than any digit *farther from* the decimal point.

PRACTICE

Answer questions 1–4.

1 The number 1 is 0.01 times _?_. ______

2 The number 0.1 is 0.01 times _?_. ______

3 The number 0.01 is 0.001 times _?_. ______

4 Which number is greater, 1 or 0.01 times 80? ______

Circle the least number in each set.

5 0.1 0.0283

6 0.0089 0.13

7 1.123 0.059

Circle the least number in each set.

8 0.02 0.005 0.1

9 1.013 1.301 1.32

To round a decimal to a particular place value, look at the digit to the right of that place value. If the digit to the right is 5 or more, round up. If the digit to the right is 4 or less, round down.

Examples:

0.0916 rounded to the tenths place is 0.1.
0.0916 to the hundredths place is 0.09.
0.0916 to the thousandths place is 0.092.

Round each decimal number to the indicated place.

10 Round 1.2341 to the hundredths place. ______

11 Round 0.0057 to the hundredths place. ______

12 Round 3.015 to the tenths place. ______

13 Round 12.1956 to the hundredths place. ______

14 Round 5.0103 to the thousandths place. ______

Adding Decimals

To add the numbers **4.3, 2,** and **0.093,** start by writing the numbers in column form with the decimal points lined up. For the whole number **2,** write a decimal point just to the right of the ones place.

41.3 2. + 0.093	Line up the decimal points.	41.300 2.000 + 0.093 43.393 ← Add the columns, from right to left.

Write a decimal point in the sum. In an addition problem, the decimal points should line up vertically.

Write a zero in each empty column. Then add the columns, from right to left. When you are finished adding each column, write a decimal point in your answer. All the decimal points should line up vertically.

PRACTICE

Add and then check your answers using rounded values. Be sure to start with the problem written in column form, and write a decimal point after each whole number.

1 32 + 0.01 = ______

2 0.06 + 0.17 = ______

3 1.5 + 0.023 = ______

4 1.03 + 0.56 = ______

5 0.09 + 0.015 = ______

6 0.035 + 0.29 + 1.1 = ______

7 7.15 + 0.25 = ______

8 9.08 + 0.15 = ______

9 15.129 + 8.08 = ______

10 0.031 + 10 + 2.2 = ______

Subtracting Decimals

To subtract decimals, start by writing the numbers in column form, with the decimal points lined up. Write a zero in each empty column, then subtract. The decimal point in the difference should line up with the other decimal points.

```
                                    6 10
  12.7    Line up the              12.70  <--- Write a zero in any empty column. Then
- 0.16    decimal points.         - 0.16       subtract, from right to left.
                                   12.54
                                     ^--- Write a decimal point in the difference. In a
                                          subtraction problem, the decimal points
                                          should line up vertically.
```

To subtract from a row of zeros, a shortcut is to write a digit one less than the first nonzero digit from the right. Next, write a ten over the zero farthest to the right and write 9 over the other zeros. Then you are ready to subtract.

```
                          3 9 9 9 10
   24.0000               24.0000
 - 0.0028              - 0.0028
                         23.9972
```

PRACTICE

Subtract and then check your answers using addition.

1 3.1 – 0.1 = ______

2 What is 0.765 – 0.003?

3 1.5 – 0.015 = ______

4 0.5 – 0.07 = ______

5 What is 1.01 subtracted from 4.25?

6 What is 0.025 subtracted from 15.6?

7 What is $35.19 – $4.07?

8 What is 47 cents subtracted from $1.15?

9 What is 0.03 subtracted from 0.264?

10 What is 0.8 subtracted from 9?

11 What is 3 cents subtracted from $15.61?

12 What is 0.017 subtracted from 5?

Multiplying Decimals

When you multiply decimals, at first you ignore the decimal points. Write the problem with the numbers lined up at the right. After you multiply, count the number of digits to the right of each decimal point. That is the **number of decimal places** for each number.

The number of decimal places in the product must equal the *sum* of the decimal places in the numbers you multiplied.

$$\begin{array}{r} \scriptstyle 1\ 2 \\ 2.24 \\ \times\ 0.5 \\ \hline 1.120 \end{array}$$

For multiplication, start by lining up the numbers at the right. Ignore the decimal points until the end.

Count the number of decimal places: 2.24 has 2 places; 0.5 has 1 place. Since 2 + 1 = 3, the product should have 3 decimal places.

PRACTICE

Finish each problem by writing the answer with a decimal point.

1 $\begin{array}{r} 0.14 \\ \times\ 2 \\ \hline .28 \end{array}$

2 $\begin{array}{r} 0.7 \\ \times\ 0.3 \\ \hline 21 \end{array}$

3 $\begin{array}{r} 1.5 \\ \times\ 5 \\ \hline 75 \end{array}$

4 $\begin{array}{r} 2.01 \\ \times\ 0.3 \\ \hline 603 \end{array}$

5 $\begin{array}{r} 21.2 \\ \times\ 0.4 \\ \hline 848 \end{array}$

6 $\begin{array}{r} 0.92 \\ \times\ 0.12 \\ \hline 184 \\ 92 \\ \hline 1104 \end{array}$

7 3 × 0.4 = ______

8 7 × 0.16 = ______

9 12 × 1.2 = ______

10 0.5 × 0.6 = ______

11 0.21 × 0.5 = ______

> "of" *and* "×"
>
> $\frac{1}{2}$ **of 10** means $\frac{1}{2}$ × **10**
>
> **0.4 of 20** means **0.4** × **20**

12 What is 0.02 of 20?

13 What is 0.14 of 100?

14 What is 5 tenths of $2.50?

15 Currants cost 3 dollars a pound. How much would 0.5 pounds cost? (Write your answer with two digits to the right of the decimal point.)

In the problem below, 0.15 has two decimal places and 0.03 has two decimal places. That means the product must have 2 + 2 = 4 decimal places. To write the product, you have to write two zeros *to the left* of the nonzero digits.

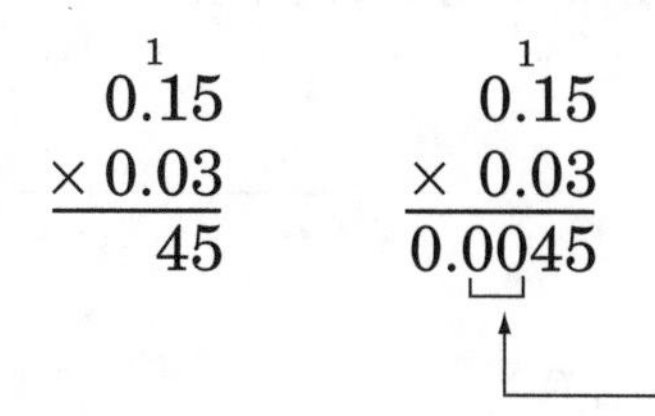

The product should have 4 decimal places. Write two zeros to the left of "45".

Multiply and then check your work using estimation.

16 $0.2 \times 0.3 =$ ______

17 $0.03 \times 0.4 =$ ______

18 0.025×0.05

19 $0.19 \times 0.2 =$ ______

20 0.62×0.003

21 $0.005 \times 0.7 =$ ______

22 0.04×0.012

23 What is 0.3 of 1.2 ounces?

24 What is 2 hundredths of 12.06?

25 What is 3 hundredths of 0.2?

26 What is one tenth of 0.5?

27 What is 3 tenths of 0.09?

28 What is 25 hundredths (or 25 percent) of 40?

29 A miner wants to divide 1.2 ounces of gold equally among 10 people. What is one tenth of 1.2 ounces?

30 What is 25 hundredths (or 25 percent) of 60?

31 What is 0.05 (or 5 percent) of 15?

32 Sales tax is 8 percent. What is 0.08 of $16.00? (Be sure your answer has the appropriate number of zeros for an amount of money.)

33 Sales tax is 12 percent. What is 0.12 of 40 dollars? (Be sure your answer has the appropriate number of zeros for an amount of money.)

Dividing a Decimal by a Whole Number

To divide a decimal by a whole number, start by writing a decimal point above the decimal point in the dividend. Then divide as usual.

Problem: 6.04 ÷ 2 = ______

$$\begin{array}{r} . \\ 2\overline{)6.04} \end{array}$$

Write a decimal point directly above the decimal point inside the division bracket.

$$\begin{array}{r} 3.02 \\ 2\overline{)6.04} \end{array}$$

Then divide, writing the digits in the quotient.

Always write a zero in the quotient if a dividend digit is too small. In the problem below, both dividend digits 2 and 4 are less than the divisor 5.

$$\begin{array}{r} 0.00408 \\ 5\overline{)0.02040} \end{array}$$

PRACTICE

Find each quotient. Check your answers using multiplication.

1 $2\overline{)4.08}$

2 $3\overline{)0.189}$

3 $5\overline{)80.5}$

4 $6\overline{)\$12.24}$

5 $7\overline{)210.63}$

6 $32\overline{)3.360}$

7 $21\overline{)10.5}$

8 $9\overline{)9.207}$

9 $7\overline{)37.1}$

10 0.912 ÷ 3 = ______

11 What is 1.35 divided by 5?

12 What is 0.1053 divided by 13?

13 5.040 ÷ 40 = ______

14 What is $42.84 divided equally among 6 people?

15 What is $756.00 divided equally among 21 people?

Dividing a Decimal by a Decimal

To divide a decimal by a decimal, start by moving both decimal points until the divisor is a whole number.

$0.33\overline{)0.528}$	$33.\overline{)52.8}$	Move both decimal points until the divisor is 33, a whole number.
	$\begin{array}{r} . \\ 33.\overline{)52.8} \end{array}$	Then write a decimal point for the quotient.
	$\begin{array}{r} 1.6 \\ 33.\overline{)52.8} \\ \underline{33} \\ 198 \\ \underline{198} \\ 0 \end{array}$	Divide 33 into 52.6. The quotient is 1.6.

As another example, you know that $\frac{4.8}{0.08} = \frac{4.8 \times 100}{0.08 \times 100} = \frac{480}{8}$.

So the division problem $0.08\overline{)4.8}$ has the same answer as $008.\overline{)480.}$ or $8\overline{)480.}$

PRACTICE

Find each quotient.

1 $0.8\overline{)76.80}$

2 $0.07\overline{)0.42}$

3 $0.32\overline{)2.88}$

4 $0.3\overline{)14.88}$

5 $1.6\overline{)7.68}$

6 $0.08\overline{)1.92}$

7 What is $0.276 \div 0.12$?

8 What is $5.26 divided by $0.20?

9 What is $53.9 \div 0.07$?

10 What is $42.84 divided by 6 cents?

11 Hot dogs are $1.50 each. How many can you buy for $6.00?

12 A phone call costs $2.88. You pay $0.12 a minute. How long is the call?

Dividing a Whole Number by a Decimal

If the number *inside* the division bracket is a whole number, start by writing a decimal point at the right of that number. Then move both decimal points until the divisor is a whole number.

Work	Explanation
$0.4\overline{)28}$ $0.4\overline{)28.0}$	Write a decimal point and a zero inside the division bracket.
$4\overline{)280.}$	Move both decimal points until the divisor is 4, a whole number. Write the decimal point for the quotient.
$\begin{array}{r} 70. \\ 4\overline{)280.} \\ \underline{28} \\ 0 \\ \underline{0} \\ 0 \end{array}$	Divide 4 into 280. The quotient is 70.

PRACTICE

Find each quotient.

1 $0.9\overline{)360}$

2 $0.04\overline{)13}$

3 $1.6\overline{)96}$

4 $0.012\overline{)144}$

5 $0.03\overline{)189}$

6 $81 \div 0.03 =$ ______

7 $768 \div 2.4 =$ ______

8 $574 \div 0.2 =$ ______

9 What is 168 divided by 1.4?

10 What is 648 divided by 0.27?

11 Joe has $20.00. Gas is $1.30 per gallon. How many full gallons can he buy?

12 A box of spices costs 5 dollars. It contains 12.5 ounces. How much does the spice cost per ounce?

Solving Mixed Word Problems

Below, circle the letter for the number sentence or expression that shows how to solve the problem.

1 A 4-ounce bottle of perfume costs \$26.56. How much does the perfume cost per ounce?

A $\$26.56 + 4 =$ ______
B $\$26.56 - 4 =$ ______
C $\$26.56 \times 4 =$ ______
D $\$26.56 \div 4 =$ ______

2 Zora bought 8 bottles of detergent. Each contains 14.3 ounces. How many ounces did she buy in all?

F $14.3 + 8 =$ ______
G $14.3 - 8 =$ ______
H $14.3 \times 8 =$ ______
J $14.3 \div 8 =$ ______

3 An uncooked roast weighs 5.3 pounds. After it is cooked, it weighs 3.7 pounds. How much did the roast shrink while cooking?

A $5.3 + 3.7 =$ ______
B $5.3 - 3.7 =$ ______
C $5.3 \times 3.7 =$ ______
D $5.3 \div 3.7 =$ ______

4 A book costs \$4.15 plus 9 cents per dollar in sales tax. How much will the book cost altogether?

F $(\$4.15 \times 0.09) + \$4.15$
G $\$4.15 + \$0.09$
H $\$4.15 \times \$0.09$
J $(\$4.15 \div \$0.09) + \$4.15$

Solve each word problem below. Use estimation to check your answers.

5 When Dan began his road trip, the odometer on his car read 1,256.7 miles. When he finished the trip, it read 1,814.9 miles. How far did Dan drive?

6 Renee finished a 7.8-mile hike in 2 hours. How fast did she walk?

7 French fries cost \$0.45 per order. Hamburgers cost \$3.59 each. How much would a hamburger and 3 orders of fries cost?

8 A 12-ounce bag of potato chips costs \$3.72. A 10-ounce bag of barbecue chips costs \$2.90. How much more do the potato chips cost per ounce than the barbeque chips? (*Hint:* This is a 2-step problem.)

Decimals Skills Practice

Circle the letter for the correct answer to each problem. Try crossing out unreasonable answers before you start to work on each problem.

1 $9.15 + 0.058 =$ ______

A 9.73
B 9.108
C 9.208
D 0.973
E None of these

2 $0.103 - 0.07 =$ ______

F 0.133
G 0.033
H 0.005
J 0.05
K None of these

3 $15.42 \times 0.003 =$ ______

A 46.26
B 4.626
C 0.4626
D 0.04626
E None of these

4 $2.1\overline{)462}$

F 220
G 2.2
H 0.022
J 22
K None of these

5 $2.79 \div 31 =$ ______

A 9
B 0.9
C 0.09
D 90
E None of these

6 $95 - 0.015 =$ ______

F 80
G 95.085
H 94.085
J 94.095
K None of these

Use the following information to do Numbers 7 through 10.

At the end of the eighth grade, Lin was 3.6 feet tall. Over the next 2 years he grew 1.2 feet.

7 How tall was Lin at the end of the tenth grade?

A 4.4 feet
B 4.8 feet
C 2.4 feet
D 4.32 feet

8 On average, how much did Lin grow per month during the ninth grade and tenth grades?

F 1 foot
G 0.05 feet
H 0.5 feet
J 1.2 feet

9 One meter is about 3.3 feet. To find Lin's height in meters at the end of the eighth grade, which should you do?

A Add 3.6 and 3.3.
B Multiply 3.6 by 3.3.
C Subtract 3.3 from 3.3.
D Divide 3.6 by 3.3.

10 In eighth grade, the tallest boy in Lin's class was 6.2 feet tall. What was the difference between his height and Lin's height?

F 2.6 feet
G 3.6 feet
H 3.4 feet
J 5.0 feet

11

$$\begin{array}{r} 9.82 \\ +\ 0.3 \\ \hline \end{array}$$

A 9.83
B 10.12
C 11.12
D 12.82
E None of these

12

$23\overline{)0.161}$

F 7
G 0.7
H 0.071
J 0.007
K None of these

13

$$\begin{array}{r} 18 \\ -\ 0.25 \\ \hline \end{array}$$

A 15.5
B 1.55
C 17.75
D 18.75
E None of these

14 $4.35 × 11 = ______

F $47.85
G $8.70
H $87.00
J $478.50
K None of these

15 2.04 − 0.009 = ______

A 2.031
B 2.31
C 2.0391
D 2.0301
E None of these

16

$$\begin{array}{r} \$15.80 \\ -\ 9.65 \\ \hline \end{array}$$

F $7.15
G $5.15
H $6.25
J $6.15
K None of these

17 0.95 + 0.79 = ______

A 1.64
B 1.74
C 0.0174
D 0.164
E None of these

18 Which of these number sentences could be used to find one-tenth of forty-six?

F 10 × 46 = ______
G 46 ÷ 0.1 = ______
H 46 × 0.01 = ______
J 46 × 0.1 = ______

19 What is 6.075 in words?

A six hundred and seventy-five thousandths
B six and seventy-five hundredths
C six and seventy-five thousands
D six and seventy-five thousandths

20 Which set of decimal numbers is in order from least to greatest?

F 0.19 0.85 1.003 0.091
G 0.091 0.19 0.85 1.003
H 1.003 0.19 0.85 0.091
J 0.19 1.003 0.85 0.091

21 A decimal numbers starts with "0." and then uses each of the digits 9, 3, 0, and 1 exactly once. What is the least possible value for the decimal number?

A 0.0931
B 0.0139
C 0.1390
D 0.9301

22 What is 47.1093 rounded to the nearest hundredth?

F 47.1
G 47.2
H 47.109
J 47.11

Fractions

Numerators and Denominators

A number such as $\frac{1}{2}$ (one-half) is called a **fraction.** A fraction represents a part of something. If you divide a cake into 8 equal pieces, each piece is $\frac{1}{8}$(one-eighth) of the cake.

In a fraction, the number on the bottom tells how many parts make up one whole object. The bottom number is called the **denominator.** The top number in a fraction is called the **numerator.** It refers to the number of parts that are shown.

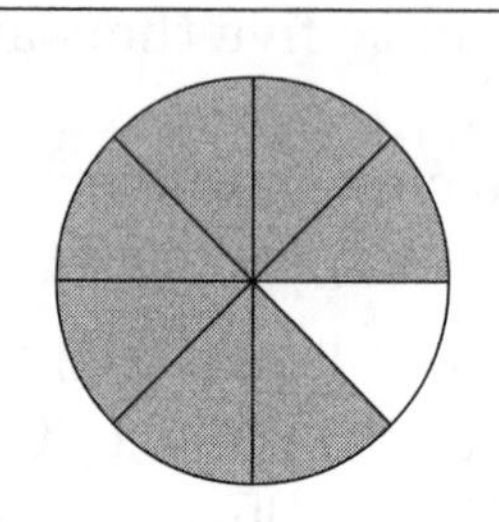

In this figure, $\frac{7}{8}$ is shaded.

It is important to understand the relationship between fractions, multiplication, and division.

Examples:

$\frac{1}{3}$ of 312 $= \frac{1}{3} \times 312 = \frac{312}{3}$ *or* $3\overline{)312}$

$\frac{1}{6}$ of 600 $= \frac{1}{6} \times 600 = \frac{600}{6}$ *or* $6\overline{)600}$

Also, $\frac{11}{20}$ can be written as $11 \div 20$ *or* $20\overline{)11}$.

PRACTICE

Write a fraction for each situation below.

1 There are 100 cents in a dollar. What fraction of a dollar is 17 cents? ______

2 There are 12 inches in a foot. What fraction of a foot is 5 inches? ______

3 Darlene has 59 pages to type. So far, she has typed 23 pages. What fraction of the job is finished? ______

4 $3 \div 8$ ______

5 $12 \div 8$ ______

Write each fraction as a division problem. Then find the quotient.

Example: $\frac{1}{2}$ of 12: <u>12</u> ÷ <u>2</u> or <u>6</u>

6 $\frac{1}{4}$ of 400: ______ ÷ ______ or ______

7 $\frac{1}{3}$ of 39: ______ ÷ ______ or ______

8 $\frac{1}{5}$ of 100: ______ ÷ ______ or ______

9 $\frac{1}{6}$ of 36: ______ ÷ ______ or ______

10 $\frac{1}{8}$ of 40: ______ ÷ ______ or ______

11 $\frac{1}{2}$ of 300: ______ ÷ ______ or ______

Comparing Fractions

The numerator (the top number) refers to the number of parts shown or shaded. If two fractions have the same *denominator* (bottom number), the fraction with the greater numerator is the greater fraction.

The denominator tells how many parts are in one object. If two fractions have the same *numerator,* the fraction with the least denominator is the greater fraction.

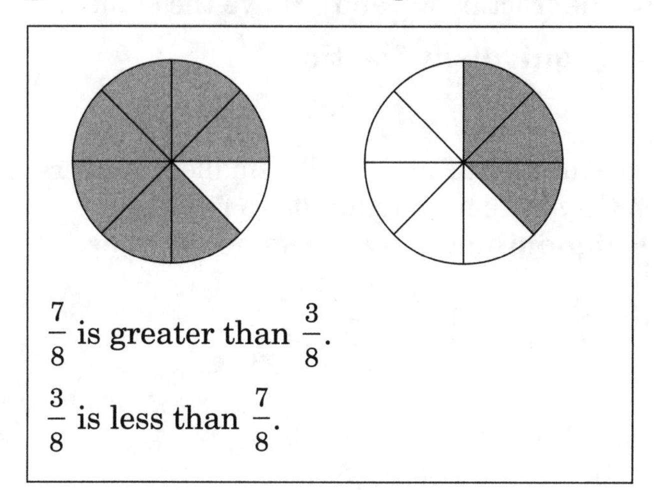

$\frac{7}{8}$ is greater than $\frac{3}{8}$.

$\frac{3}{8}$ is less than $\frac{7}{8}$.

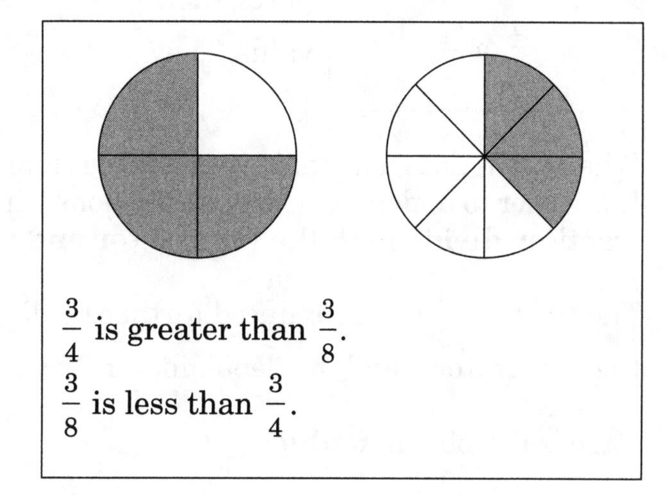

$\frac{3}{4}$ is greater than $\frac{3}{8}$.

$\frac{3}{8}$ is less than $\frac{3}{4}$.

PRACTICE

In each box below, write a “>” symbol (“is greater than”) or a “<” symbol (“is less than”).

1

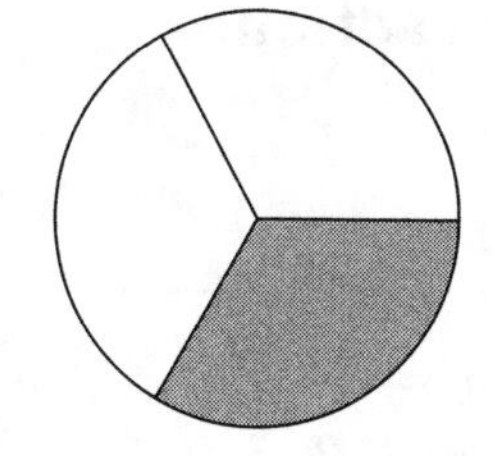

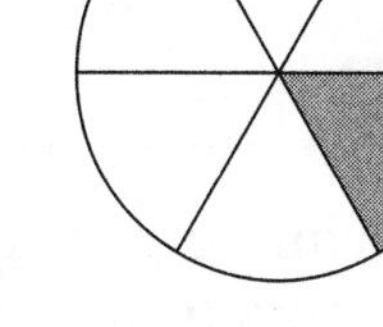

$\frac{1}{3}$ ☐ $\frac{1}{6}$

2

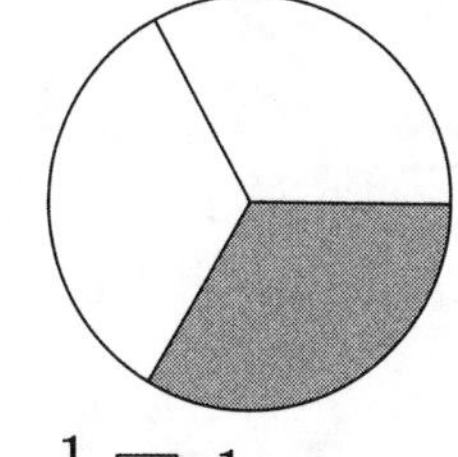

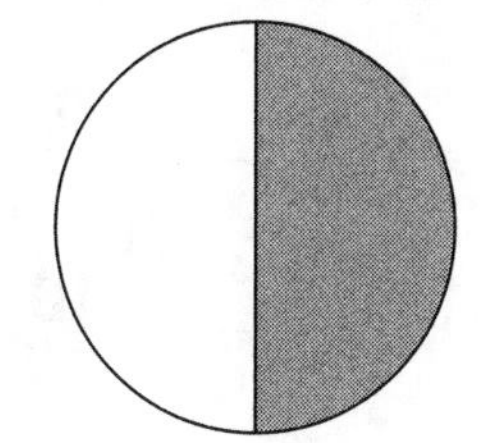

$\frac{1}{3}$ ☐ $\frac{1}{2}$

3 $\frac{1}{5}$ ☐ $\frac{1}{7}$

Arrange each set of fractions from least to greatest.

4 $\frac{1}{7}$ $\frac{1}{5}$ $\frac{1}{9}$

5 $\frac{2}{3}$ $\frac{2}{5}$ $\frac{2}{9}$

6 $\frac{1}{8}$ $\frac{1}{7}$ $\frac{1}{9}$

7 $\frac{3}{7}$ $\frac{5}{7}$ $\frac{1}{7}$

8 $\frac{1}{5}$ $\frac{2}{5}$ $\frac{2}{3}$

9 $\frac{2}{9}$ $\frac{4}{5}$ $\frac{4}{9}$

10 $\frac{1}{6}$ $\frac{1}{3}$ $\frac{1}{4}$

Reducing a Fraction to Simplest Terms

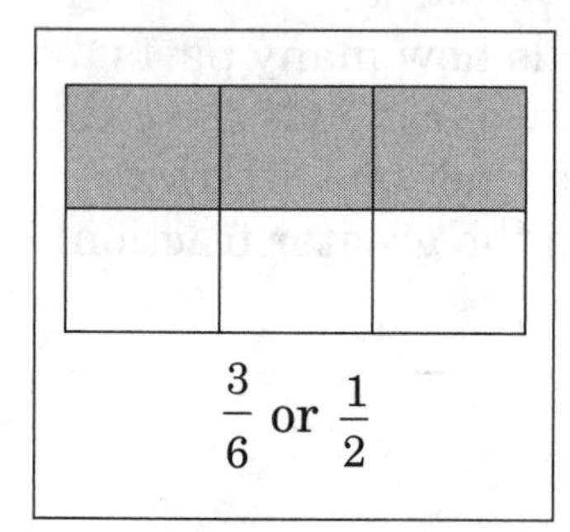
$\frac{3}{6}$ or $\frac{1}{2}$

There are several ways to name the fraction that represents this figure. If you think of the figure as being divided into six parts, then $\frac{3}{6}$ is shaded. If you think of it as being divided into two parts, then $\frac{1}{2}$ is shaded. The fractions $\frac{3}{6}$ and $\frac{1}{2}$ have the same value. They are called are **equivalent fractions.**

When you reduce a fraction, you change it into an equivalent fraction with smaller numbers. It is easier to understand and use fractions when they have been reduced. **To reduce a fraction, divide both the numerator and the denominator by the same number.**

The fraction $\frac{3}{9}$ can be reduced further by dividing the numerator 3 and the denominator 9 by 3:

$$\frac{3}{9} = \frac{3 \div 3}{9 \div 3} = \frac{1}{3}$$

Here is a problem: **Reduce $\frac{12}{30}$.**

Both the numerator 12 and the denominator 30 can be evenly divided by 2:

$$\frac{12}{30} = \frac{12 \div 2}{30 \div 2} = \frac{6}{15}$$

The new numerator 6 and denominator 15 can be evenly divided by 3:

$$\frac{6}{15} = \frac{6 \div 3}{15 \div 3} = \frac{2}{5}$$

No number can evenly divide both 2 and 5, so the fraction $\frac{2}{5}$ cannot be reduced any more. It is in **lowest** or **simplest terms.** This is called **reducing a fraction.**

PRACTICE

Reduce each fraction to simplest terms.

Hint: Any proper fraction with a prime number in the denominator is in simplest terms. The first eight prime numbers are 2, 3, 5, 7, 11, 13, 17, and 19. So fractions such as $\frac{1}{5}$, $\frac{3}{5}$, and so on are in simplest form.

1 $\frac{4}{20}$ ______

2 $\frac{45}{150}$ ______

3 $\frac{12}{30}$ ______

4 $\frac{44}{88}$ ______

5 $\frac{50}{200}$ ______

6 $\frac{24}{80}$ ______

7 $\frac{16}{36}$ ______

8 $\frac{25}{125}$ ______

9 $\frac{20}{640}$ ______

10 $\frac{50}{350}$ ______

11 $\frac{5}{25}$ ______

12 $\frac{7}{28}$ ______

Solve each problem. Reduce all fractions to simplest terms.

13 Reginald got a $500.00 bonus. The company deducted $200.00 from his bonus for taxes. What fraction of Reginald's bonus was deducted for taxes?

14 Laura Lee finds a sweater she likes for $48.80. It is marked $\frac{1}{2}$ off. What is $\frac{1}{2}$ of $48.80?

15 There were 1,200 auto thefts committed this year in Lakeview. Six hundred of the thefts were committed during the summer. What fraction of 1,200 is 600?

16 Claude is saving to buy a leather couch for $1,200. He has saved $300 toward the price of the couch. What fraction of the money has he saved so far?

17 Leanne is trying to choose between two party dresses. Both normally cost $120, but one is marked $\frac{1}{4}$ off and the other is marked $40.00 off. Which of the two dresses is less expensive?

18 There is an 84-dollar purse on sale for $\frac{1}{4}$ off. How much would you save if you bought the purse on sale?

19 The booster's club raised $450.00 selling raffle tickets. One person bought $150.00 of raffles. What fraction of the raffle-ticket sales did that person buy?

20 Ricky is getting married. The 6 other members of his work crew are buying him a gift. Each person will pay $\frac{1}{6}$ of the cost. If they buy a $126-dollar gift, how much will each person pay?

21 Rachelle's gas tank holds 24 gallons. She needs to buy $\frac{1}{4}$ of a tank of gas. How many gallons of gas will she buy?

22 Joseph has a bakery. He is thinking about marking his cakes $\frac{1}{3}$ off. If he does that, how much would a $12.00 cake be marked down?

23 Tony, his sister, and his brother are buying their mom a $240 coat. Tony's sister is going to pay $\frac{1}{3}$ of the cost. What is $\frac{1}{3}$ of $240?

24 Tony's brother is going to pay $\frac{1}{4}$ of the cost of the coat, and Tony will pay the rest of the cost of the coat. Which person will pay the most?

Fractions Equal to 1 and Fractions Greater than 1

So far, you have worked with fractions such as $\frac{4}{5}$ or $\frac{1}{3}$, where the top number is less than the bottom number. These fractions are called **proper fractions.** If the top number in a fraction is equal to or greater than the bottom number, such as $\frac{5}{4}$, $\frac{7}{6}$, and $\frac{9}{9}$, the fraction is called an **improper fraction.**

In an improper fraction, the numerator is equal to or greater than the denominator. An improper fraction represents 1 unit or a number greater than 1 unit.

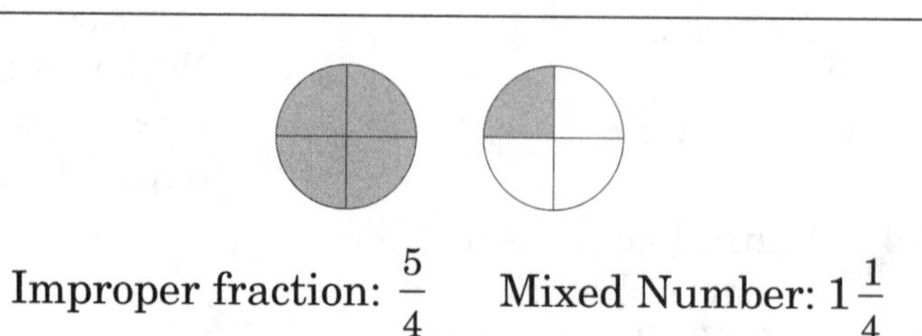

Improper fraction: $\frac{5}{4}$ Mixed Number: $1\frac{1}{4}$

Here are some other numbers that are greater than one: $3\frac{3}{4}$, $6\frac{1}{2}$, $1\frac{1}{4}$, and so on. Each of these is a **mixed number.** A mixed number is the sum of a whole number and a proper fraction.

To change an improper fraction to a whole number, start by dividing the top number by the bottom number. If there is no remainder, your answer is a whole number.

$$\frac{8}{2} = 8 \div 2 = 4$$

If there is a remainder, then your answer is a mixed number. For the fraction, the numerator is the remainder and the denominator is the denominator of the original improper fraction.

$$\frac{7}{5} = 7 \div 5 = 1 \text{ r } 2 \text{ so } \frac{7}{5} = 1\frac{2}{5}$$

PRACTICE

In each box below, write a ">" sign, a "<" sign, or an "=" sign to compare each fraction to the whole number 1.

1 $\frac{2}{3}$ ☐ 1

2 $\frac{5}{4}$ ☐ 1

3 $\frac{6}{7}$ ☐ 1

4 $\frac{13}{25}$ ☐ 1

5 $\frac{50}{50}$ ☐ 1

6 $\frac{2}{5}$ ☐ 1

7 $\frac{6}{6}$ ☐ 1

8 $1\frac{2}{3}$ ☐ 1

9 $1\frac{3}{4}$ ☐ 1

10 $2\frac{4}{5}$ ☐ 1

Write each improper fraction as a whole number or as a mixed number. Reduce all proper fractions to simplest terms.

11 $\frac{4}{3}$ = ______

12 $\frac{6}{3}$ = ______

13 $\frac{5}{4}$ = ______

14 $\frac{50}{10}$ = ______

15 $\frac{95}{40}$ = ______

16 $\frac{12}{8}$ = ______

17 $\frac{7}{5}$ = ______

18 $\frac{30}{25}$ = ______

19 $\frac{9}{5}$ = ______

20 $\frac{18}{5}$ = ______

Changing Between Decimals and Fractions

Since a quarter of a dollar is the same as 25¢, you know that $\frac{1}{4} = 0.25$. Also, since 10¢ is one tenth of a dollar, you know that $0.10 = \frac{1}{10}$. Any number that can be written as a decimal can also be written as a fraction.

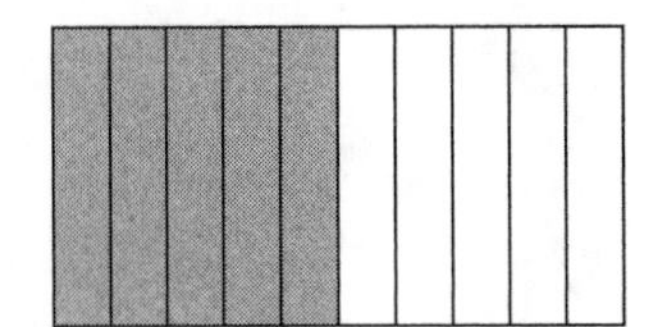

Each section is one tenth or 0.1.

In the figure above, five-tenths is shaded.

$$0.5 = \frac{5}{10} = \frac{1}{2}$$

To write a decimal as a fraction, say or write the decimal in words. Then write it as a fraction and simplify the fraction.

$0.1 = \text{one tenth} = \frac{1}{10}$

$0.15 = \text{fifteen hundredths}$

$= \frac{15}{100} = \frac{5}{5} \times \frac{3}{20} = \frac{3}{20}$

$0.035 = \text{thirty-five thousandths}$

$= \frac{35}{1000} = \frac{5}{5} \times \frac{7}{200} = \frac{7}{200}$

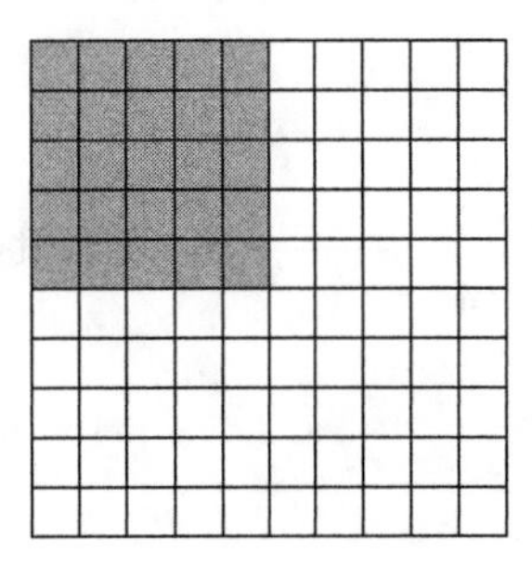

Each section is one-hundredth or 0.01.

Above, $\frac{1}{4}$ of the figure is shaded. $\frac{1}{4} = \frac{25}{100}$

To write a fraction as a decimal, divide the numerator (top number) by the denominator (bottom number). You will have to add zeros inside the bracket.

$$\begin{array}{r} 0.25 \\ 4\overline{)1.00} \\ \underline{8} \\ 20 \\ \underline{20} \\ 0 \end{array}$$

PRACTICE

Write each decimal below as a fraction. Then reduce the fraction to simplest terms.

1 0.4 ________

2 0.45 ________

3 4.2 ________

4 0.08 ________

5 0.8 ________

Write each fraction below as a decimal.

6 $\frac{1}{5}$ ________

7 $\frac{1}{20}$ ________

8 $\frac{3}{4}$ ________

Adding Fractions and Mixed Numbers

To add two fractions that have the same denominator, use that same denominator and add the two numerators. Reduce the sum if it is not in lowest terms.

$\frac{3}{8} + \frac{1}{8} = \frac{3+1}{8}$ ← Add the two numerators.

$= \frac{4}{8}$

$= \frac{4 \div 4}{8 \div 4}$ ← Reduce the fraction.

$= \frac{1}{2}$

To add two mixed numbers, add the fractions and simplify the sum. Then add the whole numbers.

If the sum of the fractions is a mixed number, carry the whole number to the column of whole numbers.

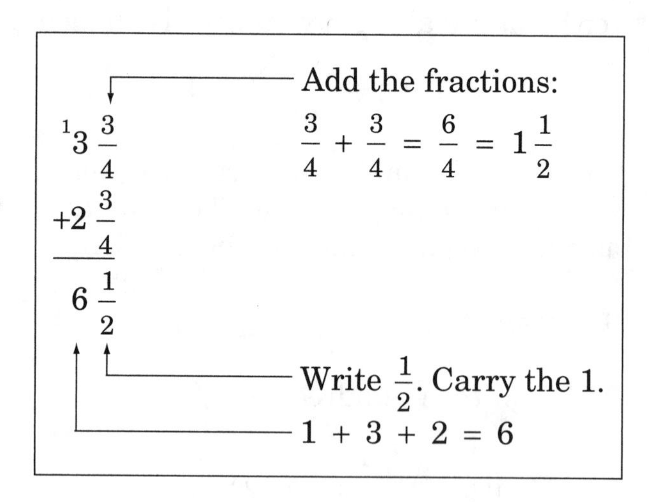

PRACTICE

Add each pair of fractions or mixed numbers below. If your answer is an improper fraction or a mixed number, simplify the sum.

1 $\frac{2}{9} + \frac{5}{9} =$ ______

2 $\frac{3}{10} + \frac{7}{10} =$ ______

3 $\frac{4}{9} + \frac{8}{9} =$ ______

4 $\frac{8}{15} + \frac{2}{15} =$ ______

5 $\frac{5}{12} + \frac{1}{12} =$ ______

6 $\frac{8}{25} + \frac{2}{25} =$ ______

7 $\frac{2}{21} + \frac{5}{21} =$ ______

8 $3\frac{3}{8} + 2\frac{5}{8} =$ ______

9 $\frac{7}{16} + \frac{7}{16} =$ ______

10 $\frac{6}{10} + \frac{9}{10} =$ ______

11 $1\frac{2}{6} + 1\frac{1}{6} =$ ______

12 $\frac{3}{18} + \frac{7}{18} =$ ______

13 $1\frac{1}{4} + \frac{1}{4} =$ ______

14 $3\frac{1}{3} + 2\frac{1}{3} =$ ______

15 $6\frac{3}{5} + \frac{1}{5} =$ ______

Subtracting Fractions and Mixed Numbers

In the subtraction problem below, the fractions have the same denominator. To subtract the fractions, use that same denominator and subtract the numerators. Reduce the difference if it is not in lowest terms.

$$\begin{array}{r} 8\frac{3}{8} \\ -\,2\frac{1}{8} \\ \hline 6\frac{1}{4} \end{array}$$

$\frac{3}{8} - \frac{1}{8} = \frac{2}{8} = \frac{1}{4}$

$8 - 2 = 6$

To subtract mixed numbers, start by subtracting the fractions. Then subtract the whole numbers.

PRACTICE

Subtract each pair of fractions or mixed numbers below. Reduce your answers to simplest terms. *Remember:* Zero over any number is zero, so $\frac{6}{7} - \frac{6}{7} = \frac{0}{7} = 0$.

1 $\frac{9}{15} - \frac{4}{15} =$ ______

2 $\frac{4}{12} - \frac{3}{12} =$ ______

3 $3\frac{3}{7} - 2\frac{1}{7} =$ ______

4 $\frac{7}{8} - \frac{3}{8} =$ ______

5 $1\frac{7}{10} - \frac{5}{10} =$ ______

6 $\frac{5}{6} - \frac{1}{6} =$ ______

7 $\frac{12}{15} - \frac{3}{15} =$ ______

8 $2\frac{2}{9} - \frac{2}{9} =$ ______

9 $3\frac{7}{15} - \frac{2}{15} =$ ______

10 $\frac{19}{20} - \frac{4}{20} =$ ______

11 $10\frac{3}{8} - 6\frac{1}{8} =$ ______

12 $5\frac{7}{10} - 1\frac{3}{10} =$ ______

13 $3\frac{1}{4} - 2\frac{1}{4} =$ ______

Borrowing to Subtract Mixed Numbers

The subtraction problem below, $5\frac{1}{3} - 1\frac{2}{3} =$ ________, involves borrowing. Rewrite $5\frac{1}{3}$ as $4\frac{4}{3}$. Then subtract the fractions and subtract the whole numbers.

Example 1: $5\frac{1}{3} - 1\frac{2}{3} =$ ________

$\frac{1}{3} < \frac{2}{3}$, so you need to borrow.

$5 + \frac{1}{3} = 4 + \frac{3}{3} + \frac{1}{3} = 4\frac{4}{3}$

$$\begin{array}{r} 4\frac{4}{3} \\ -1\frac{2}{3} \\ \hline 3\frac{2}{3} \end{array}$$

$\frac{4}{3} - \frac{2}{3} = \frac{2}{3}$

$4 - 1 = 3$

Example 2: $15\frac{1}{4} - 3\frac{3}{4} =$ ________

$\frac{1}{4} < \frac{3}{4}$, so you need to borrow.

$15 + \frac{1}{4} = 14 + \frac{4}{4} + \frac{1}{4} = 14\frac{5}{4}$

$$\begin{array}{r} 14\frac{5}{4} \\ -\ 3\frac{3}{4} \\ \hline 11\frac{1}{2} \end{array}$$

$\frac{5}{4} - \frac{3}{4} = \frac{2}{4} = \frac{1}{2}$

$4 - 3 = 1$

$1 - 0 = 1$

PRACTICE

Subtract each pair of fractions below. Reduce answers to simplest terms.

1 $1\frac{1}{5} - \frac{4}{5} =$ ________

2 $2\frac{1}{4} - \frac{3}{4} =$ ________

3 $5\frac{3}{7} - 2\frac{5}{7} =$ ________

4 $2\frac{3}{8} - \frac{4}{8} =$ ________

5 $3\frac{1}{3} - 1\frac{2}{3} =$ ________

6 $5\frac{1}{6} - 2\frac{5}{6} =$ ________

7 $12\frac{2}{5} - 3\frac{4}{5} =$ ________

8 $1\frac{7}{30} - \frac{11}{30} =$ ________

9 $1\frac{1}{4} - \frac{2}{4} =$ ________

10 $2\frac{1}{10} - \frac{3}{10} =$ ________

11 $5\frac{3}{5} - \frac{4}{5} =$ ________

Adding and Subtracting Unlike Fractions

Two fractions are "like" if they have the same denominator. They are "unlike" if they have different denominators. To add or subtract unlike fractions, start by rewriting them with a common denominator.

Example 1

Rewrite $\frac{2}{3}$ and $\frac{5}{8}$ as like fractions.

Solution

A common denominator for $\frac{2}{3}$ and $\frac{5}{8}$ is twenty-fourths, because 24 is a **common multiple** of 3 and 8.

$\frac{2}{3} = \frac{2 \times 8}{3 \times 8} = \frac{16}{24}$ $\quad \frac{5}{8} = \frac{5 \times 3}{8 \times 3} = \frac{15}{24}$

The two fractions are $\frac{16}{24}$ and $\frac{15}{24}$.

Finding the denominator 24 for the fractions $\frac{2}{3}$ and $\frac{5}{8}$ is called **finding the common denominator** of the fractions.

Example 2

$\frac{5}{6} + \frac{11}{12} = \frac{10}{12} + \frac{11}{12}$ Rewrite the fractions.

$= \frac{10 + 11}{12}$ Add the fractions.

$= \frac{21}{12} = \frac{12}{12} + \frac{9}{12}$ Simplify the sum.

$= 1\frac{9}{12} = 1\frac{3}{4}$ Reduce.

Example 3

$\frac{2}{3} - \frac{5}{8} = \frac{16}{24} - \frac{15}{24}$ Rewrite as like fractions.

$= \frac{16 - 15}{24}$ Subtract the fractions.

$= \frac{1}{24}$

PRACTICE

Write a number in each box so the two fractions are equivalent. (This is called "raising the fraction to higher terms.")

1 $\frac{1}{3} = \frac{\square}{9}$

2 $\frac{3}{5} = \frac{\square}{10}$

3 $\frac{3}{4} = \frac{\square}{16}$

4 $\frac{1}{6} = \frac{\square}{12}$

5 $\frac{2}{5} = \frac{\square}{15}$

6 $\frac{3}{4} = \frac{\square}{12}$

7 $\frac{3}{8} = \frac{\square}{16}$

8 $\frac{1}{3} = \frac{\square}{21}$

9 $\frac{1}{2} = \frac{\square}{10}$

10 $\frac{1}{2} = \frac{\square}{8}$

11 $\frac{1}{4} = \frac{\square}{20}$

12 $\frac{2}{9} = \frac{\square}{27}$

The **least common denominator** for two fractions is the least number that can be used as a denominator for the two fractions. For example, the two fractions $\frac{1}{2}$ and $\frac{3}{5}$ have common denominators of 10, 20, 30, 40, and so on. The denominator 10 is the least of the common denominators. As another example, the fractions $\frac{1}{3}$ and $\frac{4}{9}$ have common denominators of 9, 18, 27, and so on, but 9 is the least common denominator.

Find the least common denominator for each pair of fractions.

13 $\frac{1}{5}$ and $\frac{1}{10}$ ______

14 $\frac{1}{3}$ and $\frac{1}{5}$ ______

15 $\frac{1}{6}$ and $\frac{1}{9}$ ______

16 $\frac{1}{7}$ and $\frac{1}{2}$ ______

17 $\frac{3}{4}$ and $\frac{5}{6}$ ______

18 $\frac{4}{9}$ and $\frac{1}{5}$ ______

19 $\frac{2}{3}$ and $\frac{3}{7}$ ______

Raise each pair of fractions so they are *like fractions.*

20 $\frac{1}{3}$ and $\frac{1}{4}$ ______

21 $\frac{1}{5}$ and $\frac{1}{15}$ ______

22 $\frac{1}{4}$ and $\frac{1}{5}$ ______

23 $\frac{2}{3}$ and $\frac{2}{7}$ ______

24 $\frac{2}{3}$ and $\frac{5}{6}$ ______

25 $\frac{5}{6}$ and $\frac{5}{8}$ ______

26 $\frac{3}{5}$ and $\frac{2}{7}$ ______

In each problem below, rewrite the fractions so they are *like fractions.* Then add or subtract as indicated. Be sure to write your answers in simplest terms.

27 $\frac{1}{3} + \frac{1}{4}$

28 $\frac{3}{5} - \frac{1}{10}$

29 $\frac{2}{3} + \frac{1}{6}$

30 $\frac{2}{5} - \frac{3}{15}$

31 $\frac{1}{2} - \frac{1}{4}$

32 $\frac{1}{3} + \frac{2}{9}$

33 $\frac{2}{3} + \frac{1}{5}$

34 $\frac{3}{6} - \frac{1}{9} =$ ______

35 $\frac{2}{3} - \frac{1}{4} =$ ______

36 $\frac{5}{6} - \frac{1}{3} =$ ______

37 $\frac{1}{4} - \frac{1}{8} =$ ______

38 $\frac{3}{8} + \frac{1}{6} =$ ______

Multiplying Fractions

To multiply two fractions, multiply the numerators together. Then multiply the denominators together.

$$\frac{2}{4} \times \frac{3}{6} = \frac{2 \times 3}{4 \times 6} = \frac{6}{24}$$ Multiply the numerator and multiply the denominators.

$$\frac{6}{24} = \frac{6 \div 6}{24 \div 6} = \frac{1}{4}$$ Reduce the fraction.

To multiply a fraction by a whole number, rewrite the whole number as an improper fraction with a denominator of 1. Then multiply the fractions.

$$\frac{2}{3} \times 6 = \frac{2}{3} \times \frac{6}{1}$$ Rewrite 6 as $\frac{6}{1}$.

$$= \frac{2 \times 6}{3 \times 1}$$ Multiply the fractions.

$$= \frac{12}{3}$$

$$= 4$$ Simplify the fraction.

PRACTICE

Multiply the fractions below. Reduce your answers to simplest terms.

1 $\frac{1}{4} \times \frac{3}{5} =$ ______

2 $\frac{1}{3} \times \frac{2}{5} =$ ______

3 $\frac{2}{3} \times \frac{2}{3} =$ ______

4 $\frac{1}{5} \times \frac{3}{5} =$ ______

5 $9 \times \frac{2}{6} =$ ______

6 $\frac{3}{5} \times \frac{2}{3} =$ ______

7 $\frac{1}{8} \times 3 =$ ______

8 $\frac{4}{9} \times \frac{1}{5} =$ ______

9 $7 \times \frac{1}{5} =$ ______

10 $\frac{1}{2} \times \frac{1}{2} \times \frac{2}{3} =$ ______

11 $\frac{1}{2} \times \frac{2}{5} \times \frac{3}{4} =$ ______

12 $\frac{1}{7} \times \frac{3}{5} \times 3 =$ ______

13 $\frac{1}{3} \times \frac{2}{3} \times 5 =$ ______

14 What is $\frac{4}{5}$ of 10?

15 What is $\frac{2}{3}$ of 15?

16 Elaine is making half a recipe for rice. The recipe calls for $\frac{1}{4}$ cup of butter. What is $\frac{1}{2}$ of $\frac{1}{4}$ cup?

17 At Don's office, $\frac{2}{9}$ of the people smoke. There are 45 workers in the office. How many of them smoke?

Canceling Before You Multiply

To multiply $\frac{2}{5}$ times $\frac{3}{4}$, notice that the factors in the numerator and denominator have a common factor 2:

$$\frac{2}{5} \times \frac{3}{4} = \frac{2 \times 3}{5 \times 4} = \underline{\ ?\ }$$

A shortcut is to reduce the fractions before you multiply. This shortcut is called **canceling.**

To cancel, look for a common factor in the numerator and the denominator. Divide both numbers by the common factor. Then multiply as usual.

$$\frac{2}{5} \times \frac{3}{4} = \frac{\overset{1}{\cancel{2}} \times 3}{5 \times \underset{2}{\cancel{4}}}$$

Divide 2 into the numerator and divide 2 into the denominator.

$$= \frac{3}{10}$$

Then multiply the numerators and multiply the denominators.

In this problem, you can cancel two times.

$$\frac{3}{4} \times \frac{2}{3} \times \frac{1}{2} = \frac{\overset{1}{\cancel{3}} \times \overset{1}{\cancel{2}} \times 1}{\underset{2}{\cancel{4}} \times \underset{1}{\cancel{3}} \times 2} = \frac{1}{4}$$

Cancel by the factors 2 and 3.

PRACTICE

Simplify the problems below by canceling. Then multiply.

1 $\frac{2}{9} \times \frac{3}{5} =$ ______

2 $\frac{1}{6} \times \frac{2}{3} =$ ______

3 $\frac{3}{10} \times \frac{5}{9} =$ ______

4 $\frac{3}{4} \times \frac{1}{9} \times \frac{2}{5} =$ ______

5 $\frac{2}{3} \times \frac{3}{4} \times \frac{4}{5} =$ ______

6 $6 \times \frac{1}{10} =$ ______

7 $3 \times \frac{1}{3} \times \frac{1}{5} =$ ______

8 $\frac{2}{7} \times \frac{1}{2} =$ ______

9 What is $\frac{1}{3}$ of 33?

10 How much meat is in $\frac{2}{3}$ of $\frac{3}{4}$ of a pound of hamburger?

11 What is $\frac{4}{5}$ of $\frac{1}{2}$?

Dividing a Fraction by a Fraction

The numbers 4 and $\frac{1}{4}$ are called **reciprocals** of each other. To find the reciprocal of a fraction, turn the fraction "upside down."

You know that $12 \div 4$ is the same as $12 \times \frac{1}{4}$. A rule is that dividing by a number is the same as multiplying by the reciprocal of that number. You can use the same rule for fractions: Dividing by a fraction is the same as multiplying by the reciprocal of that fraction.

Example 1

$\frac{7}{2} \div \frac{1}{4} = \frac{7}{2} \times \frac{4}{1}$ Dividing by $\frac{1}{4}$ is the same as multiplying by $\frac{4}{1}$.

$= \frac{7 \times 4}{2 \times 1}$ Multiply and simplify.

$= \frac{28}{2}$

$= 14$

Example 2

$\frac{2}{3} \div \frac{5}{8} = \frac{2}{3} \times \frac{8}{5}$ Dividing by $\frac{5}{8}$ is the same as multiplying by $\frac{8}{5}$.

$= \frac{2 \times 8}{3 \times 5}$ Multiply and simplify.

$= \frac{16}{15}$

$= \frac{15}{15} + \frac{1}{15}$

$= 1\frac{1}{15}$

PRACTICE

Solve each division problem below. Be sure to reduce each answer to simplest terms.

1 $\frac{3}{4} \div \frac{1}{2} =$ ______

2 $\frac{3}{5} \div \frac{2}{5} =$ ______

3 $\frac{8}{9} \div \frac{2}{9} =$ ______

4 $\frac{4}{9} \div \frac{2}{3} =$ ______

5 $\frac{2}{3} \div \frac{5}{4} =$ ______

6 $\frac{9}{10} \div \frac{3}{8} =$ ______

7 Sara has $\frac{3}{4}$ of a yard of ribbon. She wants to divide it into $\frac{1}{3}$ yard strips. How many complete strips will she make?

8 $\frac{1}{5} \div \frac{1}{4} =$ ______

9 $\frac{1}{2} \div \frac{1}{6} =$ ______

10 Cole bought $3\frac{1}{4}$ gallons of juice for a party. He will put it into pitchers that hold $\frac{1}{2}$ gallon each. How many full pitchers will he have?

Dividing with Fractions and Whole Numbers

If a division problem has a fraction and a whole number, start by rewriting the whole number as a fraction with a denominator of 1.

Example 1

$4 \div \frac{7}{8} = \frac{4}{1} \div \frac{7}{8}$ Rewrite 4 as $\frac{4}{1}$.

$= \frac{4}{1} \times \frac{8}{7}$ Dividing by $\frac{7}{8}$ is the same as multiplying by $\frac{8}{7}$.

$= \frac{4 \times 8}{1 \times 7}$ Multiply and simplify.

$= \frac{32}{7}$

$= \frac{28}{7} + \frac{4}{7}$

$= 4\frac{4}{7}$

Example 2

$\frac{2}{3} \div 5 = \frac{2}{3} \div \frac{5}{1}$ Rewrite 5 as $\frac{5}{1}$.

$= \frac{2}{3} \times \frac{1}{5}$ Dividing by $\frac{5}{1}$ is the same as multiplying by $\frac{1}{5}$.

$= \frac{2 \times 1}{3 \times 5}$ Multiply and simplify.

$= \frac{2}{15}$

PRACTICE

Solve each division problem below. Be sure to reduce each answer to simplest terms.

1 $12 \div \frac{2}{3} =$ ______

2 $6 \div \frac{3}{4} =$ ______

3 $9 \div \frac{3}{4} =$ ______

4 $\frac{4}{7} \div 4 =$ ______

5 $\frac{1}{3} \div 2 =$ ______

6 $\frac{2}{3} \div 15 =$ ______

7 $\frac{3}{5} \div 9 =$ ______

8 $\frac{4}{7} \div 12 =$ ______

9 Wayne is putting in 5 yards of fencing. He wants a fence post every $\frac{1}{2}$ yard. How many $\frac{1}{2}$-yard sections will he have?

10 Greta has 4 cups of paint. She is pouring it into $\frac{1}{2}$-cup portions. How many portions will she have?

Solving Mixed Word Problems

For each problem below, tell whether you should add, subtract, multiply, or divide to solve the problem.

1 There are $7\frac{1}{4}$ ounces of pasta in a bag. How many ounces are in 3 bags?

A Add. **C** Multiply.
B Subtract. **D** Divide.

2 One box of pudding mix makes 3 cups. How many $\frac{1}{2}$-cup servings does it make?

F Add. **H** Multiply.
G Subtract. **J** Divide.

3 One brownie recipe calls for 2 cups of flour, $\frac{1}{4}$ cup of butter, and $\frac{1}{2}$ cup of milk. How many cups of batter will it make?

A Add. **C** Multiply.
B Subtract. **D** Divide.

4 Loni's recipe calls for 3 cups of flour. Her measuring cup only holds $\frac{3}{4}$ cup. How many measuring cups full are needed?

F Add. **H** Multiply.
G Subtract. **J** Divide.

5 Each grilled cheese sandwich costs $1\frac{1}{2}$ dollars. How many grilled cheese sandwiches can you buy for $5.00?

A Add. **C** Multiply.
B Subtract. **D** Divide.

Solve each word problem below. Reduce all answers to simplest terms.

6 To make a stuffed animal, you need $\frac{1}{4}$ yard of cloth. The store only has $\frac{1}{3}$ yard of the cloth you want. Is that enough?

7 Ronnie's book is 350 pages long. She has read 75 pages. What fraction of the book has she read so far?

8 Ron has $\frac{1}{2}$ a tank of gas. If he buys another $\frac{1}{3}$ of a tank, how much will he have?

9 The side of a carton of ice cream says it contains 12 servings and each serving is $\frac{2}{3}$ cup. How many cups of ice cream are in the carton?

10 A 36-dollar shirt is marked $\frac{1}{3}$ off. How much does the shirt cost on sale? (*Hint:* This is a 2-step problem.)

Fractions Skills Practice

Circle the letter for the correct answer to each problem. Reduce all fractions to simplest terms.

1 $1\frac{3}{5} + \frac{2}{5} =$ ______

A $1\frac{1}{5}$ **C** $2\frac{1}{5}$
B 2 **D** $2\frac{4}{5}$
E None of these

2 $\frac{7}{10} - \frac{3}{10} =$ ______

F $\frac{3}{5}$ **H** $\frac{2}{3}$
G $\frac{1}{5}$ **J** $\frac{2}{5}$
K None of these

3 $5 \times \frac{2}{3} =$ ______

A $1\frac{1}{3}$ **C** $3\frac{1}{3}$
B $2\frac{1}{3}$ **D** $\frac{2}{15}$
E None of these

4 $\frac{3}{5} \div \frac{2}{5} =$ ______

F $\frac{2}{3}$
G $1\frac{1}{2}$
H $\frac{6}{25}$
J 1
K None of these

5 $\frac{1}{3} \div 9 =$ ______

A 3
B 27
C $\frac{1}{27}$
D $\frac{1}{9}$
E None of these

Use the following information to do numbers 6 through 8.

A school band has 480 boxes of popcorn to deliver. The booster's club has offered to deliver $\frac{1}{3}$ of the boxes. The cheerleaders will deliver $\frac{1}{4}$ of the boxes.

6 What fraction of the boxes is left for the band itself to deliver?

F $\frac{5}{7}$
G $\frac{7}{12}$
H $\frac{5}{12}$
J $\frac{2}{7}$

7 The boosters will divide their portion of the deliveries among 16 people. How many boxes must each person deliver?

A 10
B 5
C 30
D 160

8 Each box of popcorn weighs $\frac{1}{2}$ pound. Which of these number sentences could be used to figure out how many pounds of popcorn the band sold?

F $480 \times \frac{1}{2} =$ ______
G $480 \div \frac{1}{2} =$ ______
H $480 \times 2 =$ ______
J $1{,}480 \times \frac{1}{2} =$ ______

9 $9 - \frac{1}{3}$ = ______

A $9\frac{2}{3}$ C $8\frac{1}{3}$

B $9\frac{1}{3}$ D $8\frac{2}{3}$

E None of these

10 $2\frac{3}{5} + 5\frac{4}{5} =$ ______

F $8\frac{2}{5}$ H $8\frac{1}{5}$

G $7\frac{5}{6}$ J $7\frac{2}{5}$

K None of these

11 $20 \div \frac{1}{2} =$ ______

A 10 C 30

B $\frac{1}{40}$ D 40

E None of these

12 $\frac{2}{5} + \frac{3}{25} =$ ______

F 1 H $\frac{13}{25}$

G $\frac{1}{5}$ J $\frac{6}{25}$

K None of these

13 $\frac{2}{3} - \frac{1}{6} =$ ______

A $\frac{1}{6}$ C $\frac{1}{2}$

B $\frac{1}{3}$ D $\frac{2}{3}$

E None of these

14 Which of the following fractions, if any, has the same value as $\frac{4}{20}$?

F $\frac{1}{4}$ G $\frac{1}{5}$

H $\frac{1}{3}$ J None of these

15 Which of these fractions has the same value as $\frac{3}{4}$?

A $\frac{6}{8}$ C $\frac{1}{3}$

B $\frac{5}{6}$ D $\frac{4}{3}$

16 Which set of fractions is in order from least to greatest?

F $\frac{1}{3}$ $\frac{1}{5}$ $\frac{1}{10}$ $\frac{1}{12}$

G $\frac{1}{12}$ $\frac{1}{3}$ $\frac{1}{5}$ $\frac{1}{10}$

H $\frac{1}{12}$ $\frac{1}{10}$ $\frac{1}{5}$ $\frac{1}{3}$

J $\frac{1}{10}$ $\frac{1}{5}$ $\frac{1}{3}$ $\frac{1}{12}$

17 Which of these number sentences is true?

A $\frac{3}{4} > 1$ C $\frac{9}{8} > 1$

B $\frac{0}{5} = 1$ D $\frac{3}{7} > 1$

18 Which of these fractions has the same value as 0.5?

F $\frac{1}{3}$ H $\frac{3}{4}$

G $\frac{1}{2}$ J $\frac{1}{5}$

19 Seven city council members voted to rezone Oak Avenue. The other eight council members voted against rezoning it. What fraction of the council voted to rezone Oak Avenue?

A $\frac{7}{8}$ C $\frac{7}{15}$

B $\frac{8}{7}$ D $\frac{15}{7}$

Signed Numbers

Positive and Negative Numbers

Look at the number line below:

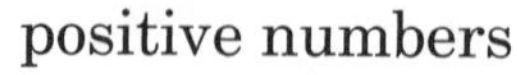

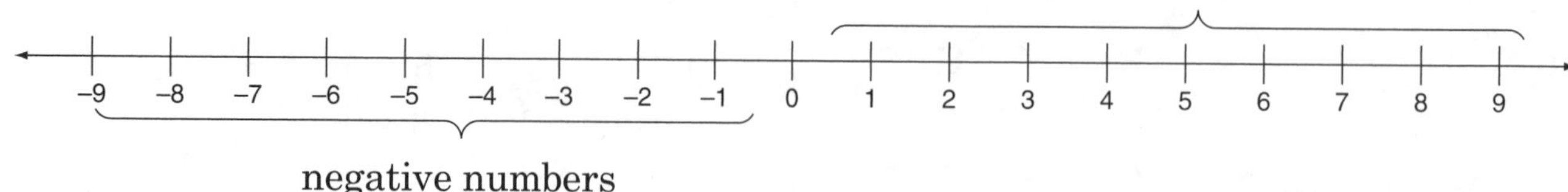

The numbers to the right of zero, such as 1, 2, $3\frac{1}{2}$, 5.7, and so on, are called the **positive numbers.** You can write the number positive five as "5" or "+5." The numbers to the left of zero, such as $-\frac{1}{2}$, –2, –8.5, and so on, are called the **negative numbers.** You always use the negative sign "–" to indicate a negative number.

You may already use negative numbers to refer to temperatures below zero. For example, two degrees below zero is written as –2°. You may also see negative numbers on bills, bank statements, and other balance sheets. The negative numbers represent money spent, while positive numbers stand for money earned.

As you move to the left on a number line, the values get less. Since –7 is to the left of –5, that means –7 is *less* than –5. Think about it this way: –7°F is colder than –5°F, or someone who owes 7 dollars has less money than someone who owes 5 dollars.

PRACTICE

Write a ">" symbol or a "<" symbol in each box to show whether the first number is greater than or less than the second number.

1 –5 □ 3

2 –2 □ –1

3 0 □ –6

4 –7 □ –3

5 –1 □ 6

6 –0.02 □ –0.025

A number's distance from zero is called its **absolute value.** The symbol for absolute value is $|\ |$.

The absolute value of –2 is 2. In symbols: $|-2| = 2$.
The absolute value of +5 is 5. In symbols, $|5| = 5$.

Write each absolute value.

7 $|3.1| =$ ______

8 $|-3| =$ ______

9 $|5| =$ ______

10 $|-5| =$ ______

11 $|-6| =$ ______

12 $|-0.25| =$ ______

Write the symbol <, =, or > in each box below.

13 $|-4|$ □ 4

14 0 □ $|-2|$

15 –9 □ $|3|$

Adding Signed Numbers

You can use a number line to illustrate addition. Find the first number on the number line. Then think of addition as a matter of moving right or left on the number line. If you are adding a positive number, you move right. If you are adding a negative number, you move left.

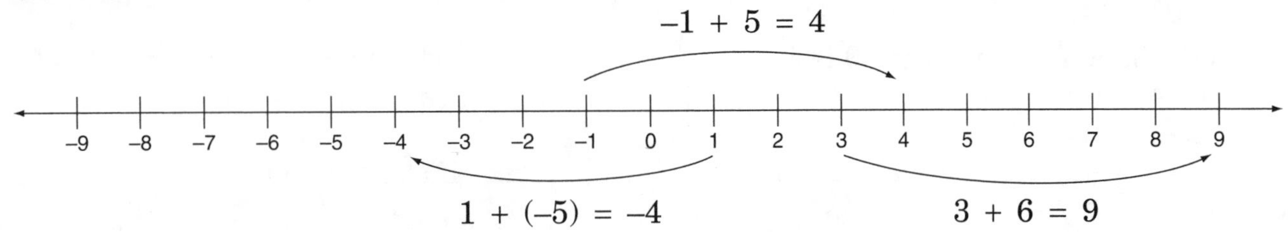

You can also use the absolute values of two signed numbers to find their sum. Here are the two rules.

If the numbers in an addition problem have the same sign (negative or positive), then:

1. Find the sum of their absolute values.
2. Give that sum their common sign.

Examples: $2 + 2 = 4$
$-3 + -4 = -7$

If the numbers in an addition problem have different signs, then:

1. Find the difference of their absolute values.
2. Give that difference the sign of the number whose absolute value is larger.

Examples: $2 + -2 = 0$
$-3 + 4 = 1$
$4 + -5 = -1$

PRACTICE

Use a number line or the rules above to fill in each blank. *Hint:* If you must add three or more numbers, add them two at a time, in any order.

1 What number is 3 to the right of zero? ______

2 What is 3 to the left of zero? ______

3 What is 5 to the right of –1? ______

4 What is 3 to the left of 2? ______

5 $-3 + (-2) =$ ______

6 $1 + (-3) =$ ______

7 $-6 + 6 =$ ______

8 $-1 + (-1) =$ ______

9 A diver's depth was –500 feet. She descended another 200 feet. At what level was she then? ______

10 The temperature was –2°F. Then it dropped 2 degrees. What was the temperature then? ______

11 What is 3 to the left of –2? ______

12 To add –4 to 5 on a number line, start at +5 and move 4 spaces to the ______.

Subtracting Signed Numbers

When you subtract one number from another, you are finding the difference between them. Therefore, you can solve a subtraction problem by finding the distance between two numbers on a number line. If you have moved left to solve the problem, give your answer a positive sign. If you have moved right, the difference is negative.

The distance from –4 to –9 is 5, so $-4 - (-9) = 5$.

The distance from 4 to –1 is 5, so $4 - (-1) = 5$.

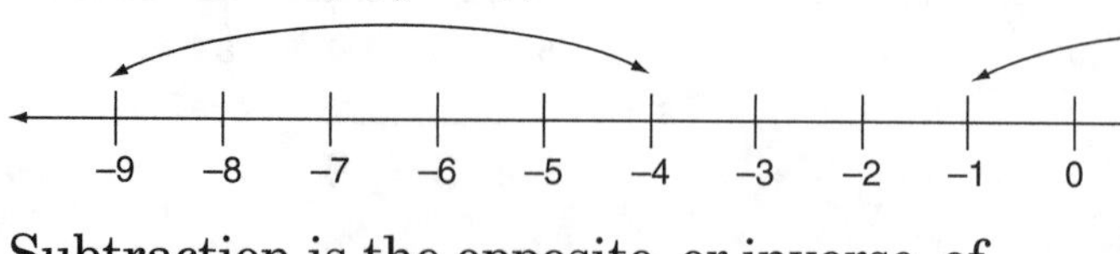

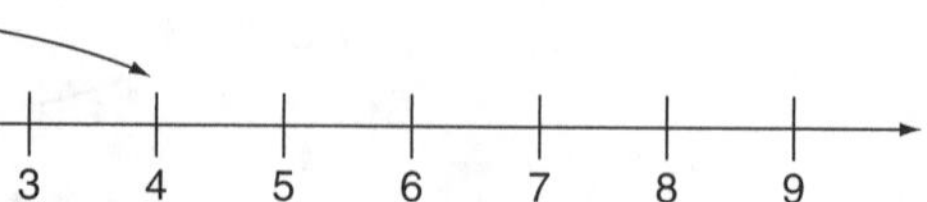

Subtraction is the opposite, or inverse, of addition. You can think of subtraction as *adding the opposite of a number.*

Examples:

$5 - 9 = 5 + (-9) = -4$

$-2 - (-4) = -2 + 4 = 2$

$-8 - (-5) = -8 + 5 = -3$

Rules for Subtracting Signed Numbers

1. Change the sign of the number being subtracted.
2. Change the subtraction sign in the problem to an addition sign.
3. Follow the rules (on page 61) to add positive and negative numbers.

PRACTICE

Solve each problem below.

1 $2 - (-3) =$ ______

2 $-3 - 2 =$ ______

3 $-4 - (-5) =$ ______

4 $2 - 5 =$ ______

5 $3 - (-3) =$ ______

6 $4 - (-3) =$ ______

7 $-7 - (-2) =$ ______

8 $-9 - 1 =$ ______

9 $0 - (-9) =$ ______

10 $10 - (-3) =$ ______

11 $8 - (-2) =$ ______

12 $-7 - (-1) =$ ______

13 $-13 - (-7) =$ ______

14 $-1 - 3 - (-1) =$ ______

15 The elevation of La Paz, Bolivia, is 12,000 feet above sea level. Death Valley, California, has an elevation of –242 feet. How much higher is La Paz than Death Valley? ______

16 Last month the balance on Wendy's credit card statement was –$135.50. This month it was –$75.50. What was the change in Wendy's balance? Write your answer with a "+" sign or a "–" sign. ______

Multiplying and Dividing Signed Numbers

If you multiply a negative number by a positive number, the product is negative. Here is an example: If you owe a credit card company one hundred dollars, your balance is –\$100.00. If your brother owes three times as much (3 × –\$100.00), his balance is –\$300. So:

$$3 \times -100 = -300$$

In a division problem, if one number is negative and the other is positive, then the quotient is negative. For example, if you owe 100 and your sister owes one-fourth as much, she owes –\$25.

So: $$-100 \div 4 = \frac{-100}{4} = -25$$

Here are rules for multiplying and dividing signed numbers:

> To multiply signed numbers, multiply their absolute values.
> To, divide signed numbers, divide their absolute values.
> Then:
> If the two numbers have the same sign, write the answer as a positive number.
> If the two numbers have different signs, write the answer as a negative number.

Examples:

$2 \times (-9) = -18$
$-2 \times 4 = -8$
$-8 \times (-5) = 40$

$-18 \div (-9) = 2$
$-8 \div 4 = -2$
$40 \div (-5) = -8$

PRACTICE

Solve each problem below.

1 $2 \times (-3) =$ ______

2 $-5 \times (-4) =$ ______

3 $-3 \times 4 =$ ______

4 $4 \times (-8) =$ ______

5 $6 \times (-5) =$ ______

6 $-8 \times (-8) =$ ______

7 $-7 \times (-5) =$ ______

8 $-2 \times 9 =$ ______

9 $6 \times (-4) =$ ______

10 $12 \div (-3) =$ ______

11 $16 \div (-2) =$ ______

12 $-20 \div (-2) =$ ______

13 $-14 \div 7 =$ ______

14 $50 \div (-25) =$ ______

If the numerator and denominator of a fraction have the same sign, the fraction is positive. If the numerator and denominator have different signs, the fraction is negative.

Simplify each fraction below.

15 $\frac{-2}{-6}$

17 $\frac{-6}{9}$

16 $\frac{4}{-8}$

18 $\frac{-4}{-12}$

Solving Mixed Word Problems

The problems below involve addition, subtraction, multiplication, or division of signed numbers. Set up and solve each problem. Use a negative sign ("–") for negative values. *Hint:* Most of these are 2-step problems.

1 Lynn had the following transactions on her credit card: –$62.91, –$9.50, –$3.10, +$3.10. What was the change to her credit card balance?

2 At noon the temperature was –6°C. Now it is –12°C. How much has the temperature changed?

3 A diver rested at –340 feet. Then she started descending, and 20 minutes later she was at –420 feet. How many feet per minute did she move?

4 Joe's restaurant has made $27,000 so far this year. But Joe's employees just went on strike, and he is losing money. Every day he adds –$1,500 to his books. What will be the total in his books after 5 days of the strike?

5 Nick just turned on his new freezer. The temperature inside is now 10°F. It is dropping 2 degrees every minute. How long will it take the temperature to reach –10°F?

6 Find the current balance for this balance sheet.

Previous balance	+$246.00
Deposit	+$12.00
Expense	–$90.00
Current balance	

Signed Numbers Skills Practice

Circle the letter for the correct answer to each problem.

1 $\frac{-12}{-24} =$ ______

A 2
B $\frac{1}{2}$
C $-\frac{1}{2}$
D –2
E None of these

2 $6 + (-7) =$ _____

F 13
G –13
H 1
J –1
K None of these

3 $-12 \div (-2) =$ _____

A 6
B –6
C 24
D –24
E None of these

4 $|5 - 9| =$ _____

F 4
G 14
H –4
J –14
K None of these

5 $9 + (-3) + (-1) =$ _____

A 5
B –5
C 7
D –7
E None of these

6 $10 - 3 =$ _____

F 13
G –13
H 7
J –7
K None of these

7 $-2 - (-9) =$ _____

A 7
B –7
C 11
D –11
E None of these

8 $|3| - |-5| =$ _____

F 2
G 8
H –2
J –8
K None of these

9 $-8 + (-3) + 4 =$ _____

A 0
B –8
C 8
D –7
E None of these

10 $\frac{80}{-10} =$ ______

F 8
G $\frac{1}{8}$
H $-\frac{1}{8}$
J –8
K None of these

11 Which set of signed numbers is in order from least to greatest?

A	–12	0	5	–30
B	–30	–12	0	5
C	0	5	–12	–30
D	–12	–30	0	5

12 What number is two less than negative thirteen?

F –11
G 11
H –15
J 15

13 Which of the following is greater than $-\frac{1}{3}$?

A $-\frac{1}{2}$
B –1
C 0
D $-\frac{2}{3}$

14 Lou had a 50-dollar credit in his credit card account. Then he spent 167 dollars. Which of these number sentences could you use to find Lou's new balance?

F –\$50.00 + (–\$167.00) = ______
G \$167.00 + \$50.00 = ______
H \$50.00 + (–\$167.00) = ______
J –\$50.00 + \$167.00 = ______

15 Which of the following expressions has the same value as –2 – (–6)?

A –2 + 6
B 2 + (–6)
C –2 – 6
D 2 – 6

16 What is the distance between –3 and 7 on a number line?

F 4
G 10
H –4
J 11

17 Which of these decimals is less than –1?

A –0.5
B 0.75
C –0.3
D –2.12

18 Which of the following has the same value as 28 – 52?

F –28 + 52
G 28 – (–52)
H –28 + (–52)
J 28 + (–52)

19 Which of the following, if any, does not equal –12?

A 12 × (–1)
B –12 × (–1)
C –12 × 1
D $\frac{-12}{1}$

Ratio and Percent

Writing Ratios

Each of the following five statements is a *ratio:*

There are 4 men *for every* woman in the group.
There is 1 murder committed *every* 5 minutes.
He scores one free throw *out of every* three he attempts.
She drove 60 miles *per* hour.
The chocolates cost $8.00 *per* pound.

A **ratio** is a comparison of two numbers. Ratios can be written in three ways:

1 to 2 **1 : 2** $\frac{1}{2}$

The third way, using a fraction, is the most useful.

The five ratios from the top of this page are shown at the right, each in fraction form. Notice that the word *per* refers to "1 unit."

$\frac{\text{4 men}}{\text{1 woman}}$ $\frac{\text{1 murder}}{\text{5 minutes}}$ $\frac{\text{1 score}}{\text{3 attempts}}$ $\frac{\text{60 miles}}{\text{1 hour}}$ $\frac{\$8.00}{\text{1 pound}}$

PRACTICE

Write each ratio below as a fraction. Use a number and a unit in both the numerator and the denominator.

1 Each inch on the map stands for 25 miles. ______

2 Ricardo earns $2,958 a month. His rent is $550. ______

3 Plant one evergreen tree for every 3 leafy trees. ______

4 Roses cost $4.00 per dozen. ______

5 Buy 3 shirts for the price of 1. ______

6 It takes me 3 hours to make 4 placemats. ______

7 The plane travels 450 miles per hour. ______

All ratios should be written as fractions reduced to lowest terms. If a ratio is an improper fraction, *do not* write it as a mixed number.

Simplify each ratio below.

8 The athletic teams have 12 men for every 8 women. ______

9 You can buy 6 ears of corn for $2.00. ______

10 We spent $340 on materials and $1,200 on labor? ______

11 They lost 6 games out of 26. ______

12 My team works for 6 days. Then we take 2.5 days off. ______

Finding a Unit Rate

In a ratio, if the number in the denominator is 1, that ratio is called a **unit rate.** Examples of unit rates are miles per gallon, meters per second, and dollars per pound.

To calculate a unit rate, such as miles per hour:

1. Put the information you have in ratio form. The word that follows *per* goes into the denominator.
2. Simplify the ratio until the bottom number is 1.

Problem: At a constant speed, it took Glenn 6 hours to drive 360 miles. How fast did he drive?

Write a ratio: $\frac{360 \text{ miles}}{6 \text{ hours}}$

Simplify it:

$$\frac{360 \text{ miles}}{6 \text{ hours}} = \frac{360 \div 6}{6 \div 6} = \frac{60 \text{ miles}}{1 \text{ hour}}$$

Answer: 60 miles per hour

PRACTICE

Find the unit rate for each problem.

1 Linc bought 40 pounds of beef for \$120. How much did he pay per pound?

2 Rita spent 50 hours making a set of ceramic tiles. Her client paid her \$800. How much did Rita make per hour?

3 For that \$800, Rita made 160 tiles. How much was she paid per tile?

4 Last year, Central Middle School spent 800,000 dollars. There are about 600 students in the school. About how much did the school spend per pupil?

5 There are 47 teachers at Central Middle School. About how many students are there per teacher?

6 Dan is going on a 21-mile hike. He wants to finish in 6 hours. How many miles must he walk each hour?

7 A 16-ounce can of Mrs. Smith's Soup costs \$1.19. A 10-ounce can of Savory Soup costs just 89 cents. Which soup costs more per ounce?

Writing Proportions

If you write an equation with one ratio equal to another ratio, such as $\frac{3}{6} = \frac{6}{12}$, you have a **proportion.** The two ratios in a proportion form **equivalent fractions.**

Proportions: $\frac{\text{3 dollars}}{\text{6 cans}} = \frac{\text{6 dollars}}{\text{12 cans}}$ $\frac{\text{10 meters}}{\text{3 seconds}} = \frac{\text{30 meters}}{\text{9 seconds}}$

In this problem you know three of the numbers in a proportion, and you need to find the fourth number.

Problem:
The cost of 12 roses is 10 dollars. How much would 24 roses cost?

Write a ratio of the two units, dollars and roses: $\frac{\text{dollars}}{\text{roses}}$.

Then write ratios using the numbers in the problem, with "?" as the unknown number. Be sure all the numerators use the numbers for dollars and all the denominators use the numbers for roses.

$$\frac{\text{dollars}}{\text{roses}} = \frac{10}{12} = \frac{?}{24}$$

You can get 24 in the denominator by multiplying 10 and 12 by 2:

$$\frac{10}{12} = \frac{10 \times 2}{12 \times 2} = \frac{20}{24}$$

Then look at the proportion $\frac{?}{24} = \frac{20}{24}$.

Answer: The cost for 24 roses is 20 dollars.

PRACTICE

Below, write each proportion in fraction form. *Remember:* Use like labels in the numerators and denominators.

1 Since 12 yards of lumber cost $40.00, then 36 yards would cost $120.00 _______

2 The electrician charges $50.00 per hour, so for 3 hours work he charges $150.00 _______

3 Apples are 2 pounds for a dollar, so one pound would cost 50 cents _______

Write each of these problems as a proportion. Use the symbol "?" for the missing number in each problem.

4 On a diagram, 1 inch represents 3 feet. What does 3 inches on the diagram represent? _______

5 In one day Lance can build 15 feet of fencing. How long would it take to build a 60-foot fence? _______

6 One piece of pie has 500 calories. How many calories would be in 3 pieces of pie? _______

For problems 7–10, tell how to solve each problem.

7 $\frac{36 \text{ inches}}{3 \text{ feet}} = \frac{? \text{ inches}}{9 \text{ feet}}$

To solve, multiply 36 inches by ______.

8 $\frac{360 \text{ feet}}{3 \text{ seconds}} = \frac{? \text{ feet}}{60 \text{ seconds}}$

To solve, multiply 360 feet by ______.

9 $\frac{2 \text{ inches}}{5 \text{ miles}} = \frac{? \text{ inches}}{60 \text{ miles}}$

To solve, multiply 2 inches by ______.

10 $\frac{4 \text{ boxes}}{2 \text{ dollars}} = \frac{? \text{ boxes}}{1 \text{ dollar}}$

To solve, *divide* 4 boxes by ______.

Solve each proportion below.

11 $\frac{2 \text{ adults}}{15 \text{ children}} = \frac{? \text{ adults}}{75 \text{ children}}$

12 $\frac{16 \text{ ounces}}{1 \text{ pound}} = \frac{? \text{ ounces}}{15 \text{ pounds}}$

13 $\frac{3 \text{ cans}}{2 \text{ dollars}} = \frac{15 \text{ cans}}{? \text{ dollars}}$

14 $\frac{70 \text{ dollars}}{5 \text{ weeks}} = \frac{140 \text{ dollars}}{? \text{ weeks}}$

15 $\frac{260 \text{ miles}}{10 \text{ gallons}} = \frac{? \text{ miles}}{5 \text{ gallons}}$

Write a proportion for each problem below. Then find the missing number in your proportion.

16 You are driving 75 miles an hour. How long will it take to drive 375 miles?

17 If 2 yards of cloth cost $10.00, how much would 5 yards of cloth cost?

18 On average, Ron makes 3 sales for every 25 calls he makes. If he makes 150 calls, how many sales can be expect to make?

19 Every 3 inches on a map stands for 350 miles. How many miles do 9 inches represent?

20 One can of juice serves 8 children. How many cans do you need for 64 children?

21 One week is 7 days. How many weeks are there in 175 days?

Cross Multiplying to Solve a Proportion

When two fractions or ratios are equal, you can write a proportion. An example is $\frac{2}{3} = \frac{8}{12}$.

Below, notice that if you **cross multiply** by finding the products 3×8 and 2×12, the two products are equal.

Proportion	**Cross Multiply**	**Cross Products**
$\frac{2}{3} = \frac{8}{12}$	$\frac{2}{3} \times \frac{8}{12}$	$3 \times 8 = 2 \times 12$ $24 = 24$

The result is true for any proportion: When you cross multiply in a proportion, the products are equal. Therefore, you can solve a proportion by following these steps:

1. Cross multiply the numbers in the proportion.

$$\frac{5 \text{ fruit bars}}{\$2} = \frac{? \text{ fruit bars}}{\$10}$$ Proportion

$$\$2 \times ? = \$50$$ Cross multiply.

2. Divide to find the value of the "?" symbol.

$$? = \frac{50}{2} = 25$$ Divide to get "?" alone.

PRACTICE

Below, circle the letter of the choice that shows how to solve the proportion.

1 $\frac{36 \text{ feet}}{2 \text{ seconds}} = \frac{? \text{ feet}}{10 \text{ seconds}}$

A Multiply 36 by 10. Then divide the product by 2.
B Multiply 36 by 2. Then divide the product by 10.
C Multiply 10 by 2. Then divide the product by 36.

2 $\frac{3 \text{ pounds}}{15 \text{ dollars}} = \frac{12 \text{ pounds}}{? \text{ dollars}}$

F Multiply 3 by 12. Then divide the product by 15.
G Multiply 12 by 15. Then divide the product by 3.
H Multiply 3 by 15. Then divide the product by 3.

Use cross multiplication to solve each proportion below.

3 $\frac{75 \text{ cents}}{15 \text{ ounces}} = \frac{? \text{ cents}}{5 \text{ ounces}}$

4 $\frac{3 \text{ birdhouses}}{8 \text{ hours}} = \frac{? \text{ birdhouses}}{40 \text{ hours}}$

5 $\frac{8 \text{ wins}}{14 \text{ games}} = \frac{? \text{ wins}}{21 \text{ games}}$

Percent

The word **percent** means "per hundred." A percent is a fraction with a written or unwritten denominator of 100.

Examples: $25\% = \frac{25}{100}$ or $\frac{1}{4}$ $\quad$ $50\% = \frac{50}{100}$ or $\frac{1}{2}$

To write a percent as a fraction, remove the percent sign and write the number as a fraction with a denominator of 100. Then simplify the fraction.

$$30\% = \frac{30}{100} = \frac{3}{10} \times \frac{10}{10} = \frac{3}{10}$$

To write a fraction as a percent, rewrite the fraction so it has a denominator of 100. Then write the numerator with a "%" symbol.

$$\frac{4}{5} = \frac{4}{5} \times \frac{20}{20} = \frac{80}{100} = 80\%$$

To write a percent as a decimal, move the decimal point two places to the left.

25% = 0.25
50% = 0.50

To write a decimal as a percent, move the decimal point two places to the right.

0.47 = 47%
0.015 = 1.5%

PRACTICE

Fill in the two missing values in each row. Reduce all fractions to simplest terms.

Percent	Fraction	Decimal
5%		
8%		
	$\frac{1}{10}$	
		0.12
		0.15
	$\frac{9}{50}$	
20%		
	$\frac{1}{4}$	
		0.3
		0.4

Percent	Fraction	Decimal
50%		
55%		
	$\frac{3}{5}$	
		0.72
		0.75
	$\frac{39}{50}$	
80%		
85%		
	$\frac{9}{10}$	

Finding a Percent of a Number

To find a percent of a given number, follow these steps:

1. Write the percent as a decimal or as a fraction.
2. Multiply that value by the given number.

Problem: A $75.00 blanket is marked 20% off. What is 20% of $75.00?

Solution 1

Write the percent as a decimal. $20\% = 0.20$

Multiply that decimal by the given number.

$$\begin{array}{r} \$75.00 \\ \times \quad 0.20 \\ \hline 15.0000 \end{array} \text{ or } \$15.00$$

Solution 2

Write the percent as a fraction. $20\% = \frac{20}{100} = \frac{1}{5}$

Multiply that fraction by the given number. $\frac{1}{5} \times 75 = \frac{1}{5} \times \frac{75}{1} = 15$ or $15

Note: When your answer is in dollars, make sure it has the correct number of decimal places. Remember: You can write or erase zeros at the right of a decimal number without changing its value.

PRACTICE

Solve each problem.

1 There are 35 people on a committee. To elect a new chairperson, 60% of the members must vote for the same person. What is 60% of 35?

2 The sales tax in Mount Clement is 14%. What is 14% of $45.00

3 A $30.00 teapot is marked down 15%. What is 15% of $30.00?

4 Helen does not want to spend any more than 30% of her income on rent. Her income is $2,300.00 a month. What is the highest monthly rent payment Helen can afford?

5 To get a high grade on his geometry test, Ricardo must get 80% of the questions correct. There are 150 questions on the test. How many questions must he get correct?

6 Len wants to leave 20% of his bill as a tip. His bill is $12.90. How much money should he leave as a tip?

Adding or Subtracting Percent

In the two-step problem below, first you find a percent of a given number. Then you subtract that value from the given number.

Problem: Find the sale price of a 35-dollar blouse marked down 25%.

Solution

First, find 25% of $35.00.

```
 $35.00
× 0.25
  17500
  7000
 8.7500 or $8.75
```

Then subtract that value from $35.00.

```
 $35.00
− 8.75
 $26.25
```

PRACTICE

Solve each of the two-step problems below.

1 A $50.00 dress is marked down 20%. How much does it cost? (Ignore the tax.)

2 A restaurant bill for 6 people is $65.00 plus a 15% tip. What is the total bill? (Ignore the tax.)

3 A magazine costs $3.50 plus 8% tax. What is the final cost to the buyer?

4 In a local election. Cox received 25% more votes than Borland. Borland received 5,420 votes. How many votes did Cox get?

5 Hillary gives 10% of her take-home pay to charity. Her yearly take-home pay is $28,000. How much is left after she makes her donations to charity?

6 An economic study reported that the average teenaged boy eats 20% more than the average teenaged girl. If a teenaged girl eats about $45.00 in food each week, how much does a teenaged boy eat each week?

Finding What Percent One Number Is of Another

To find what percent one number is of another, put the numbers in fraction form. In the fraction, write the "part" in the numerator and the "whole" in the denominator. Then reduce the fraction and write it as a percent.

Problem
Quinn spends 6 hours a week of his free time at a gym. He has only 40 hours of free time each week. What percent of his free time does Quinn spend at the gym?

Solution

First, write 6 and 40 as a fraction. $\frac{6}{40}$

Simplify the fraction. $\frac{6}{40} = \frac{6 \div 2}{40 \div 2} = \frac{3}{20}$

Write the fraction as a percent by rewriting the fraction with a denominator of 100. $\frac{3}{20} = \frac{3}{20} \times \frac{5}{5} = \frac{15}{100} = 15\%$

Answer: Quinn spends 15% of his free time at the gym.

PRACTICE

Solve each problem below.

1 Randell borrowed $9,000.00 to buy a new car. So far, he has paid $1,800.00 of his loan. What percent of the loan is paid off?

2 There are 80 questions on a social studies test. Rebecca got 48 questions right. What percent did she get right?

3 The Chin family earns $4,000 per month. On average, they spend $520 per month on groceries. What percent of their income is spent on groceries?

4 A radio originally priced at $40.00 has been marked down $16.00. By what percent has it has been marked down?

5 Valerie was making $12.00 an hour. She got a raise of $0.48 per hour. By what percent has her pay increased?

6 There are 14,000 registered voters in Waterston. In the last election, 5,600 voted. What percent of the registered voters participated in the last election?

Finding the Total When a Percent Is Given

In the problem below, the part and the percent are given. To find the total number of people, follow these steps:

1. Write the percent as a fraction.
2. Divide the given number by that fraction.

Problem: At Unified Systems, 265 workers are on strike. That is 35% of the total number of workers at the factory. How many workers are there at Unified Systems?

Solution

First, write 35% as a fraction. $\frac{35}{100}$

Simplify the fraction. $\frac{35}{100} = \frac{35 \div 5}{100 \div 5} = \frac{7}{20}$

Divide 265 by the fraction. $265 \div \frac{7}{20} = \frac{265}{1} \times \frac{20}{7} = \frac{5{,}300}{7} = 757.142$

Answer: The answer refers to a number of people, so round the answer to the nearest whole number. The total number of workers at Unified Systems is 757.

PRACTICE

Solve each problem below. Round each answer to the nearest whole number.

1 In a recent survey, about 5,240 people said they would like to change careers. That is 25% of all the people questioned. How many people were questioned?

2 Landon just got a statement saying he has paid off 30% of his car loan. So far he has paid $4,530.00. How much will he pay altogether on the loan?

3 A storm damaged Elaine's home. The insurance company charged her for 45% of the cost. They charged her $225. How much was the total amount of damage?

4 Karen and Mike have $9,000 for a down payment on a home. They know they must make a 15% down payment. What is the most expensive home they can afford?

Mixed Practice with Percent

Solve each problem below. *Hint:* Most of the problems are two-step problems.

1 There are fourteen people on the arts council. Eight of them voted to have a spring piano concert. What percent of the council voted for the concert? *(Round your answer to the nearest percent.)*

2 A $32.00 can of paint is marked down 20%. How much does it cost now?

3 Brooke paid $33.00 for a magazine subscription. The magazine publisher said that was 40% off the newsstand price. What is the newsstand price? (*Hint:* Brooke *did not* pay 40% of the newsstand price.)

4 To win an election, Maria must get 51% of the vote. If 2,942 people will vote in the election, how many votes does Maria need? (*Hint:* You cannot "round down" for this problem.)

5 Jorge bought some vases for his shop. He paid $14.00 for each vase, and he wants to mark up the price by 40%. What price should he put on each vase?

6 A blouse is marked $11.00 off its original price of $50.00. By what percent has the blouse been marked down?

7 Sales tax in Tiffany's county is 9%. How much sales tax will she pay on a $75.00 purchase?

8 Roger polled one hundred twenty people at his office about what health plan they prefer. Eighty-four of them preferred The Doctor's Choice Health Plan. What percent preferred that health plan?

Ratio and Percent Skills Practice

Circle the letter for the correct answer to each problem. Reduce all fractions to simplest terms.

1 13% of $25.00 = _____

A $325.00
B $225.00
C $3.25
D $2.25
E None of these

2 45% of ☐ = 27

F 60
G 121
H 30
J 55
K None of these

3 What percent of 80 is 20?

A 40%
B 20%
C 25%
D 60%
E None of these

4 75% of 500 = _____

F 400
G 150
H 37.50
J 375
K None of these

5 What percent of 80 is 16?

A 50%
B 20%
C 15%
D 16%
E None of these

6 Elena bought 2 pounds of sliced ham for $3.75 per pound. Sales tax is 12%. How much does Elena owe?

F $7.50
G $0.90
H $7.40
J $8.40

7 A new article says that the average American family spends 27% of its income on housing. If the average family income is $35,000 per year, how much does the average American family spend on housing?

A $7,245 per year
B $9,450 per year
C $2,450 per year
D $2,700 per year

8 Audry spends $600 per month on rent. She makes $25,000 per year. To the nearest percent, what percent of her income does she spend on rent?

F 28%
G 21%
H 20%
J 29%

9 What is the value of the symbol "?" in the proportion below?

$$\frac{5}{25} = \frac{8}{?}$$

A 200
B 40
C $\frac{40}{25}$
D 1,000

10 Riley bought 42 ounces of cereal for $12.60. How much did he pay per ounce?

F 30 cents
G 3.2 cents
H 52.9 cents
J 23 cents

11 Which of the following statements is true?

A $20\% < 15.6\%$
B $98\% < \frac{2}{3}$
C $0.3 < 25\%$
D $1 < 112\%$

12 Company rules state that Jeanette must rest 4 hours for every 6 hours she spends driving. Which of the following proportions could Jeanette use to figure out how many hours of rest she must have during a 32-hour drive?

F $\frac{4}{6} = \frac{32}{?}$

G $\frac{6}{4} = \frac{?}{32}$

H $\frac{4}{10} = \frac{?}{32}$

J $\frac{4}{6} = \frac{?}{32}$

13 Which of the following number sentences could you use to find the unknown value in the proportion below?

$$\frac{7}{20} = \frac{?}{45}$$

A $\frac{7 \times 20}{45} = \underline{?}$
B $7 \times 20 \times 45 = \underline{?}$
C $7 \times \underline{?} = 20 \times 45$
D $7 \times 45 = \underline{?} \times 20$

14 Which of the following number sentences could you use to find 32% of 67?

F $67 \div 0.32 = \underline{?}$
G $67 \times 0.32 = \underline{?}$
H $67 \times \underline{?} = 32$
J $32 \times 6.7 = \underline{?}$

15 A couch which normally costs $550.00 is marked down 20%. Which steps will give you the price of the couch on sale?

A Multiply $550 by $\frac{2}{10}$ (or $\frac{1}{5}$) and then add the product to $550.
B Multiply $550 by 0.2 and then subtract the product from $550.
C Divide $550 by 0.2 and then subtract the quotient from $550.
D Simply multiply $550 by $\frac{1}{2}$.

16 Which fraction has the same value as 25%?

F $\frac{1}{4}$ **H** $\frac{1}{25}$

G $\frac{1}{5}$ **J** $\frac{25}{10}$

Data Interpretation

Reading a Table

Tables and graphs are useful ways to organize information and show many numbers. In order to understand a table or a graph, always begin by reading the title and the headings. They explain the relationships shown in a table or graph.

To find out when Super Bowl XVI was shown, look along the row for Super Bowl XVI. Then look down the column for Date First Shown. Your answer, Jan. 24, 1982, appears where the row and column meet. →

The Most Popular Television Shows of All Time

Show	Date First Shown	Rating*	Number of Viewers
The last episode of M*A*S*H	Feb. 28, 1983	60.2%	50,150,000
Dallas (Who Shot J.R.?)	Nov. 21, 1980	53.3%	41,470,000
Roots, Part 8	Jan. 30, 1977	51.1%	36,380,000
Super Bowl XVI	Jan. 24, 1982	49.1%	40,020,000
Super Bowl XVII	Jan. 30, 1983	48.6%	40,480,000

*Percentage of U.S. households that viewed the show

PRACTICE

Use the table above to answer each question.

1 This column of numbers is taken from the table. What do these numbers show?

60.2%
53.3%
51.1%
49.1%
48.6%

A the percentage of Americans who say each show is their favorite

B the percentage of households that viewed the show when it was first shown

C the percentage of judges who voted for each show

2 When was the "Who Shot J.R.?" episode of Dallas first shown? _______

3 How many people saw the last episode of M*A*S*H when it was first shown? _______

4 Which shows had a viewer rating higher than 50%? _______

5 Which show had a viewer rating of 60.2%? _______

6 What is the oldest show mentioned in this table? _______

Using Numbers in a Table

You can use the numbers in a table to make comparisons. Start by finding each number. Then:

- You can find the **difference** between two numbers by subtracting.
- You can see how **many times larger** one number is than another by dividing.
- You can find **what fraction** one number is of another by forming a ratio.

The 10 Largest Native American Groups in the United States

Nation	Size
Cherokee	369,035
Navajo	225,298
Sioux	107,321
Chippewa	105,988
Choctaw	86,231
Pueblo	55,330
Apache	53,330
Iroquois	52,557
Lumbee	50,888
Creek	45,872

Question 1: The Chippewa Nation is about how many times larger than the Lumbee?

Solution:
Round the two numbers and divide. For this comparison, the number for the Chippewa goes inside the division bracket.

$$50{,}000\overline{)100{,}000}\quad = 2$$

Question 2: The Creek Nation is about what fraction of the size of the Chippewa?

Solution:
Round the numbers and put them in fraction form. For this comparison, the number for the Creek goes in the numerator.

$$\frac{50{,}000}{100{,}000} = \frac{1}{2}$$

PRACTICE

Use the table above to fill in the blanks.

1 What is the largest Indian group in the United States? ______

2 Which tribe is larger, the Apache or the Choctaw? ______

3 How many more Lumbee are there than Creek? ______

4 Which group is closest in size to the Chippewa? ______

5 Which group is about $\frac{2}{3}$ the size of the Cherokee? ______

6 The Sioux Nation is about what fraction of the size of the Navajo?

A $\frac{1}{5}$ **B** $\frac{1}{4}$ **C** $\frac{1}{3}$ **D** $\frac{1}{2}$

7 How many more Choctaw are there than Pueblo? ______

8 Together, the Apache and the Iroquois are about the same size as which group? ______

9 *Yes* or *No:* Does this table show how fast each group is growing? ______

Using a Price List

A menu or price list is a common type of table. To use a menu or price list, find what you want to buy. Then look for the price for that item. Do not get confused if you are looking for several prices. Write the price for each item. Then do any figuring.

Parcel Post Rate Schedule

Weight	Local	Zones 1 and 2	Zone 3	Zone 4	Zone 5	Zone 6	Zone 7	Zone 8
2 pounds	$2.56	$2.63	$2.79	$2.87	$2.95	$2.95	$2.95	$2.95
3 pounds	2.63	2.76	3.00	3.34	3.68	3.95	3.95	3.95
4 pounds	2.71	2.87	3.20	3.78	4.68	4.95	4.95	4.95
5 pounds	2.77	2.97	3.38	4.10	5.19	5.56	5.95	5.95
6 pounds	2.84	3.07	3.55	4.39	5.67	6.90	7.75	7.95
7 pounds	2.90	3.16	3.71	4.67	6.11	7.51	9.15	9.75
8 pounds	2.96	3.26	3.85	4.91	6.53	8.08	9.94	11.55
9 pounds	3.01	3.33	3.99	5.16	6.92	8.62	10.65	12.95
10 pounds	3.07	3.42	4.12	5.38	7.29	9.12	11.31	14.00

Insurance rates: $0.75 to insure packages for up to $50; $1.60 to insure packages for $50–$100; $2.50 to insure packages for $200–$300.

PRACTICE

Use the price list above to answer the following questions.

1 How much does it cost to mail a 3-pound package to zone 3? _______

2 How much more does it cost to send a 4-pound package to zone 6 than to send it to zone 2? _______

3 How much would it cost to send a 2-pound package to a local address and an 8-pound package to zone 3? _______

4 How much would it cost to send a 9-pound package to zone 4 and insure it for $250.00? _______

5 To the nearest dollar, how much would it cost to send a 2-pound package to zone 8, a 4-pound package to zone 4, and a 5-pound package to zone 6? _______

6 You have $5.00 to send an 8-pound package to zone 1. How much insurance can you afford? _______

7 You are sending a package to zone 6. How much more would it cost to send a 9-pound package rather than a 5-pound package? _______

8 How much does it cost to send a 2-pound package to zone 4? _______

9 Find the cost per pound to send a 6-lb package to zone 6? _______

10 How much would it cost to send a 3-pound package to zone 8, a 5-pound package to zone 4, and a 10-pound package to zone 2? _______

Finding the Mean, Median, and Mode

There are several different meanings for "the most typical value" in a list of values.

- One meaning of "most typical value" is the **mean** or **average.** To find an average, add all the numbers in a set. Then divide that sum by the total number of values.
- The **median** in an ordered list of numbers is the middle value. If a list has an even number of values, then there is no middle value. In this case, the median is the number *halfway between* the two middle numbers.
- The **mode** in a set of numbers is the value that appears most often. If no number appears more often than any other, then the set of values has "no mode."

Example: Here are Rob's bowling scores for his last 10 games:
151, 167, 139, 152, 163, 169, 201, 171, 168, 169

To find the mean or average:

1. Add all the scores. The sum of the ten numbers is 1,650.
2. Divide the sum by the number of scores:

$$\frac{1650}{10} = 165$$

Rob's mean score is 165.

To find the median:

1. Start by ordering the scores from least to greatest: 139, 151, 152, 163, 167, 168, 169, 169, 171, 201
2. The two middle scores are 167 and 168. The median score is halfway between these two scores.

Rob's median score is $167\frac{1}{2}$.

To find the mode:

Robert scored 169 twice. No other score appears more than once.

Rob's mode (or **modal score**) is 169.

PRACTICE

The table below shows the bowling scores of Rob's teammates. Each player bowled 4, 5, or 6 games. Fill in the missing numbers in the shaded section of the table. (If there is no mode for a set of times, write "none" or "No mode.")

Bowling Scores

Name	1	2	3	4	5	6	Median	Mode	Mean
Matt Reed	175	180	196	200	179	154	179.5	No mode	181
Ben Hanks	189	156	168	207	180	X			
Jose Ruiz	232	230	263	180	230	X			
Phil Chu	178	167	172	175	175	165			
Linda Glass	167	163	167	163	X	X		No mode	
Tim Horne	144	127	85	144	116	110			
Connie Chu	212	156	164	246	260	246			

Graphs

A graph presents information. You can find many graphs in newspapers, magazines, and reports. A graph can show a lot of information quickly and clearly.

Here are three different types of graphs:

A **circle graph** divides a circle into slices or wedges to show parts of a whole.

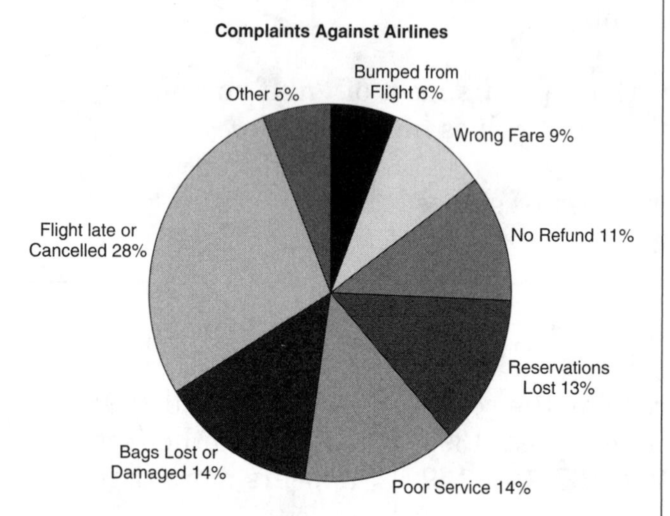

1 How many categories of complaints are shown?

A **bar graph** uses thick bars to represent numbers.

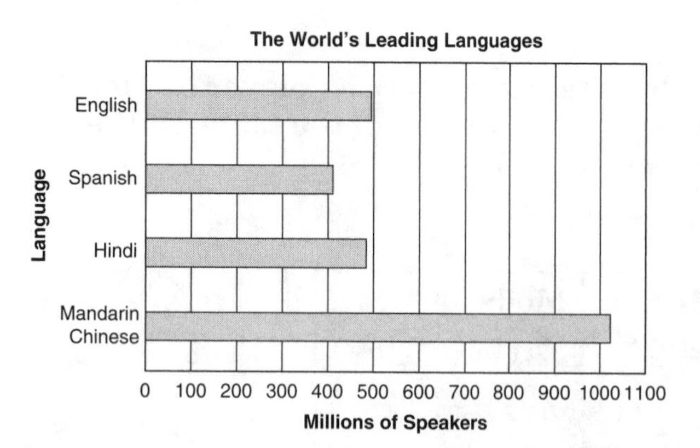

2 How many language categories are shown?

A **line graph** uses points or dots to show values. Lines connecting the points show rising or falling values.

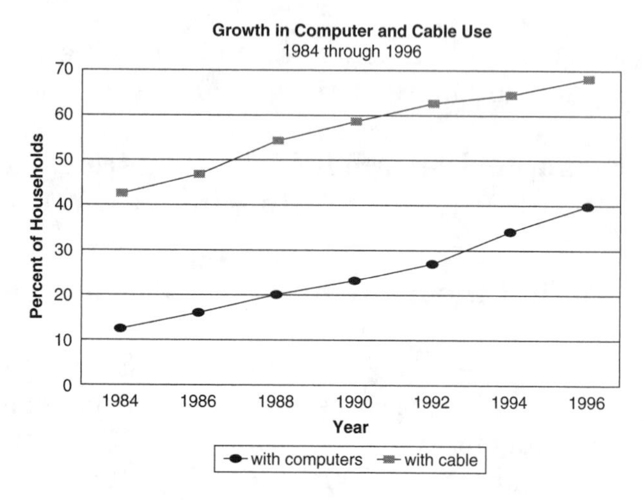

3 How many years of information are shown?

Reading a Circle Graph

A circle graph shows how a whole is divided into parts. It is divided into sections, like a pie, and each section stands for a fraction of the total. The larger the section, the larger the fraction.

In a circle graph it is easy to compare each section to the whole. In the circle graph at the right, the section for Basic Cable represents a little over one-third of what television viewers watch.

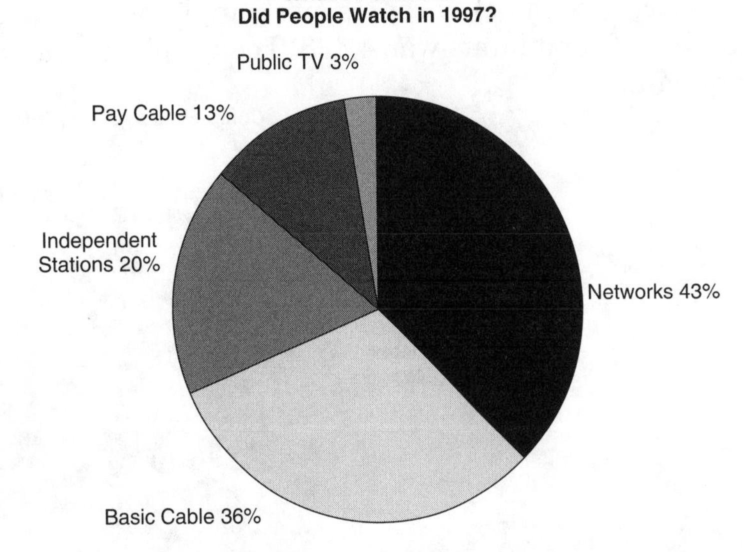

Note: As a result of multiset use and rounding of numbers, figures add to more than 100%

PRACTICE

Use the graph above to answer the following questions.

1 What type of station had the most viewers in 1997? ______

2 People viewed twice as much networks as <u>?</u>. ______

3 What type of station had about $\frac{1}{5}$ of all viewers? (Remember, to change a percent to a fraction, write it with a denominator of 100.) ______

4 Which had more viewers, the independent stations or pay cable? ______

5 About what fraction of the TV that people watched was on basic cable stations? ______

The graph below shows the types of TV that were watched in 1988. To answer Numbers 7 and 8, compare this graph to the one above.

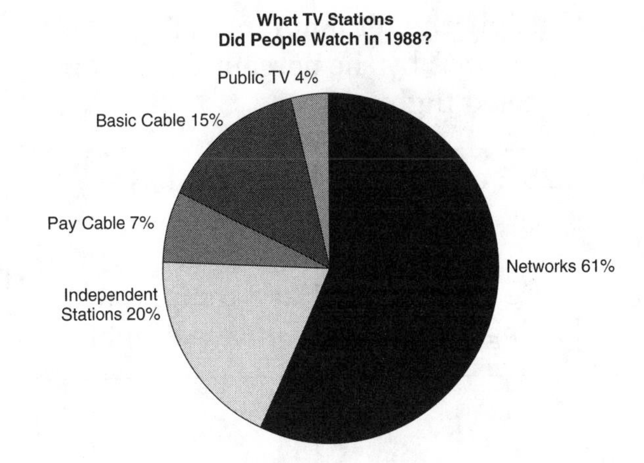

6 In which year, 1988 or 1997, did the networks have a greater percent of viewers? ______

7 Which two types of stations grew in market share between 1988 and 1997? ______

Numbers and Percents in a Circle Graph

The circle graph below represents complaints against U.S. airlines. Last year, the total number of complaints was 4,438. The graph shows categories and percents for the complaints.

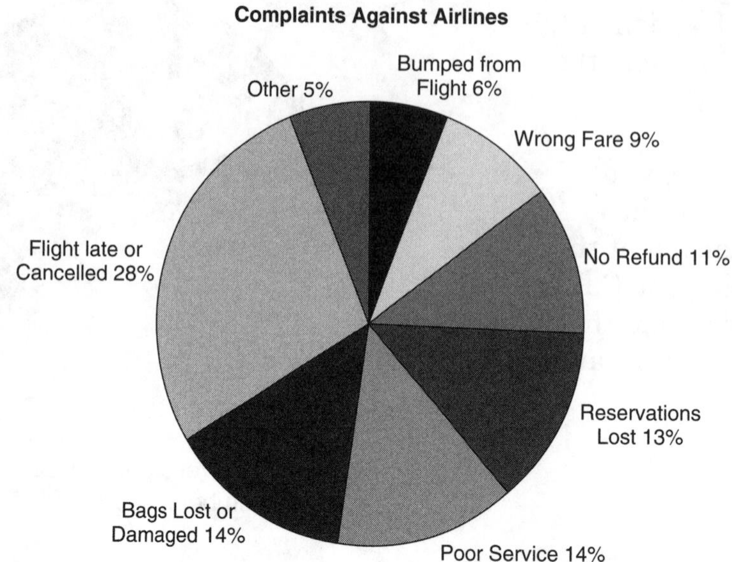

Problem 1: How many complaints were there about late or canceled flights?

Solution:
Multiply the total number of complaints by the percent for late or canceled flights.

$$\begin{aligned} 28\% \text{ of } 4{,}438 &= 0.28 \times 4{,}438 \\ &= 1242.64 \end{aligned}$$

There were about 1,243 complaints dealing with late or canceled flights.

Problem 2: How many complaints were there about lost luggage?

Solution:
Multiply the total number of complaints by the percent for lost luggage.

$$\begin{aligned} 14\% \text{ of } 4{,}438 &= 0.14 \times 4{,}438 \\ &= 621.32 \end{aligned}$$

There were about 621 complaints dealing with lost luggage.

PRACTICE

Use the graph above to answer the following questions. Round your answers to the nearest whole number.

1 How many complaints were made about poor service? ______

2 How many people complained about being charged the wrong fare? ______

3 How many people complained about lost reservations or missing refunds? ______

4 About twice as many people complained about reservations as about __?__. ______

Reading a Bar Graph

A bar graph uses thick lines, or bars, to represent values. The longer the bar, the larger the number.

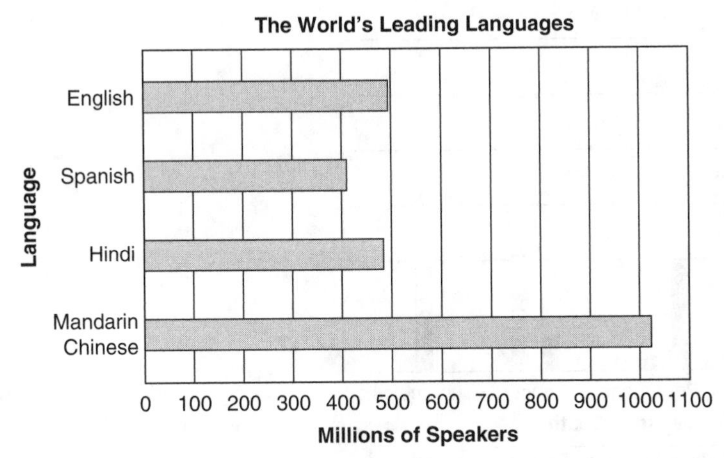

Question:
How many people speak Spanish?

Solution: Use the labels along the left side of the graph to find the bar for the subject you want—Spanish.

Using the length of that bar, find the number that corresponds to the end of the bar. It is a little more than 400.

Look at the information below the graph to identify the units used in the graph. That unit is "millions." There are over 400 million Spanish-speaking people in the world.

If the length of a bar falls between two numbers, estimate where it lies between the two numbers. For example, it might be halfway between the numbers, $\frac{1}{3}$ of the way, and so on. Then figure out the value for the end of the bar. For example, the end of the bar for Hindi lies about three-fourths of the way between 400 million and 500 million. So a good estimate for the number of people who speak Hindi is 475 million.

PRACTICE

Use the graph above to answer the following questions.

1 About how many people in the world speak English? ______

2 Which of these is the best estimate of how many people speak Mandarin Chinese?

A 1,000
B 1 billion (1,000,000,000 or 1,000 million)
C 1.2 billion
D 1.02 billion

3 About _?_ as many people speak Mandarin as speak English. ______

4 Which of the following is *not* shown on this graph?

A which language is spoken by the most people
B how many more people speak Hindi than Spanish
C how many people learn Mandarin as their first language

Using a Key or Legend

A bar graph may have several types of bars, or a line graph may have several kinds of lines. For these graphs, there is usually a box below the graph, called a **key** or **legend.** For the graph below, the key indicates that the types of bars represent data for 1900, 1950, and 1996.

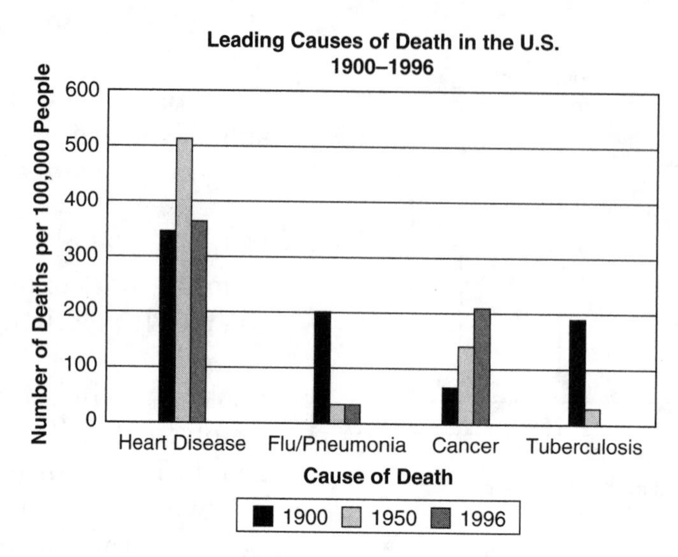

PRACTICE

Use the graph above to answer the following questions.

1 In 1996, about how many Americans out of every 100,000 died of cancer? ______

2 About how many Americans died of heart disease in 1950? (Use the correct unit.) ______

3 Which cause of death rose between 1900 and 1996? ______

4 The cause of death due to flu and pneumonia in 1950 was about what fraction of the cause of death for flu/pneumonia in 1900? ______

5 Which cause of death was about the same in 1900 and 1996? ______

6 About how much lower was the cause of death due to tuberculosis in 1950 than it was in 1900? ______

7 There is no bar shown for tuberculosis deaths in 1996. What does this suggest?

A Modern Americans cannot get tuberculosis.
B There were too few tuberculosis deaths in 1996 to show.
C Experts no longer keep track of tuberculosis deaths.

8 An average American town of 200,000 expected about how many cancer deaths in 1996? ______

9 Which cause increased the number of deaths the most between 1900 and 1950? ______

10 Which cause of death fell the most between 1950 and 1996? ______

Reading a Line Graph

A line graph uses points or dots to show values. The numbers along one side of the graph show the values of the points on the graph.

On the graph, lines between the points show whether the values are rising or falling. So line graphs show trends and changes in amounts.

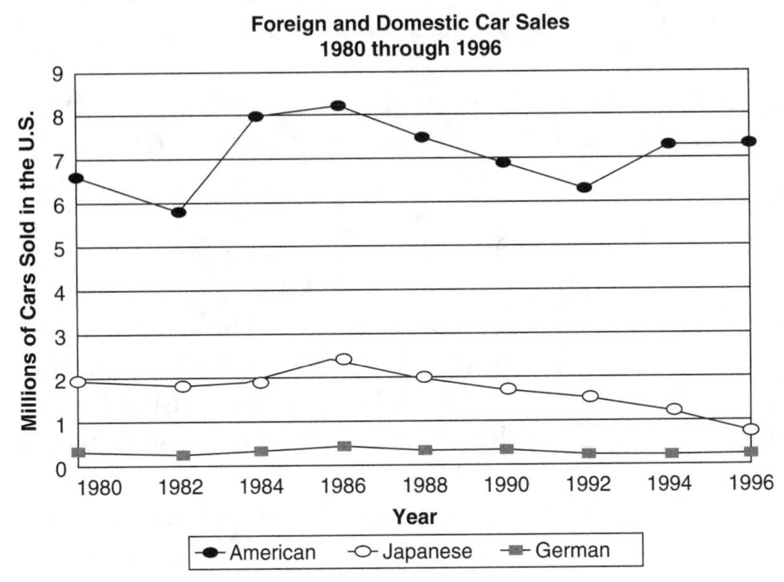

Question: During what period did American car sales rise most rapidly?

Solution: Start by looking at the key or legend. Find the line for the subject you want, which is American car sales. To find the period of most rapidly rising sales, look for the part of the line with the steepest *uphill* slope.

That part of the line begins at the dot for 1982 and ends at the dot for 1984.

Answer: 1982 through 1984

PRACTICE

Use the graph above to answer the following questions.

1 From 1986 to 1992, did American car sales rise or fall? ______

2 During which period did Japanese car sales rise most rapidly? ______

3 Japanese car sales have been falling since _?_. ______

4 About how many Japanese cars were sold here in 1980? (Be sure you use the correct units.) ______

5 About how much did American car sales rise between 1982 and 1984? ______

6 In 1986, about what was the difference between the number of Japanese and American cars sold? ______

7 In what year was there the greatest difference between the number of Japanese and American cars sold? (*Hint:* Look for the greatest distance between the two lines.) ______

Trends and Predictions

A pattern in a graph is called a **trend.** If a line graph shows values rising for several years in a row, that is a trend. From that trend you might predict that the values will continue to rise.

On a bar graph, a trend might be shown if the bars for one value are always higher than the bars for another value. From that trend you might predict that the values for the first group will continue to be higher than the values for the second group.

If the values on a graph are changing at a steady rate, you might even predict how much they will change in the future. As an example, consider the Japanese car sales from 1980 through 1996.

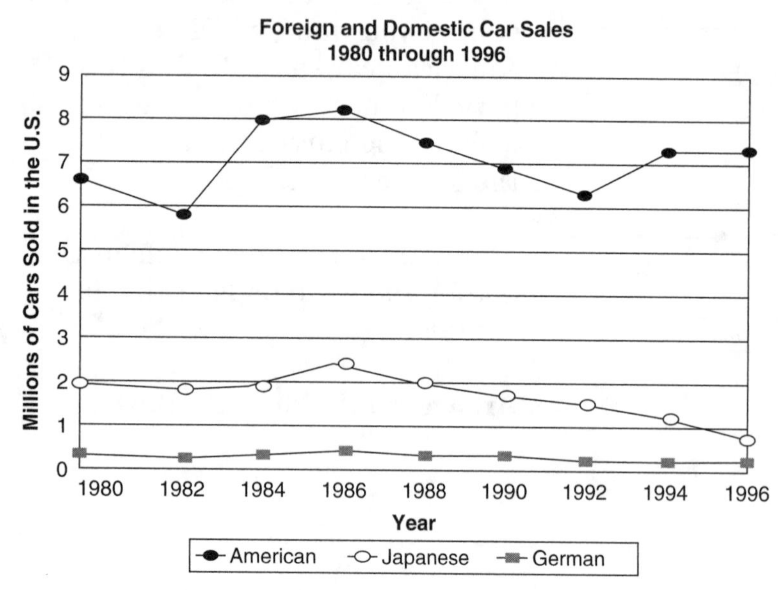

Over the ten-year period from 1986 to 1996, Japanese car sales dropped steadily. (The graph was almost a straight line.) You can use that trend to predict a value for the 1998 sales of Japanese cars.

First, estimate how much Japanese car sales dropped each two-year period. It was about $\frac{1}{3}$ of a million cars.

Since 1998 is one two-year period after 1996, you can subtract $\frac{1}{3}$ of a million cars from the number of Japanese cars sold in 1996:

$$\frac{2}{3} \text{ million} - \frac{1}{3} \text{ million} = \frac{1}{3} \text{ million}$$

PRACTICE

Use the graph above to answer the following questions.

1 Which of these statements describes the trend in German car sales?

- **A** They have been rising steadily since 1980.
- **B** They have remained about the same since 1980.
- **C** They have declined drastically since 1986.

2 Which of these statements best describes how Japanese car sales changed over the years shown?

- **F** There was a steady decline over all the years shown.
- **G** They rose when American car sales rose and fell when American car sales fell.
- **H** They rose from 1982 until 1986, then steadily declined.

What trends can you identify in the graph of "Leading Causes of Death"? Use the graph to answer questions 3 – 7.

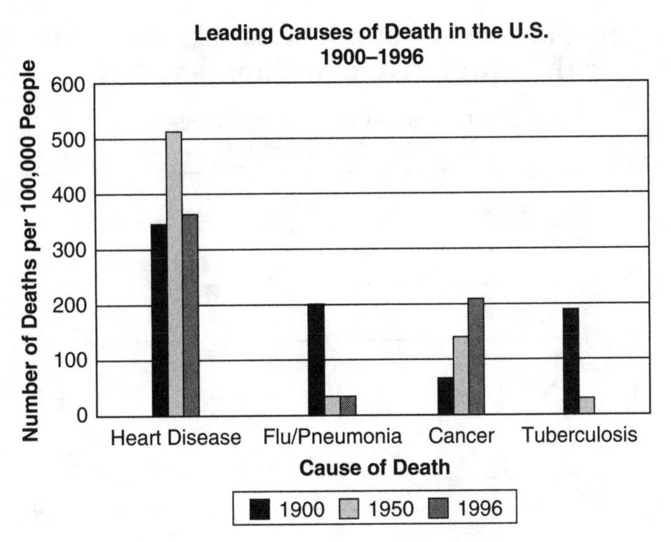

3 If the trends shown on this graph continue, what will be the number of deaths from cancer in the year 2050?

A about 100 per 100,000 people
B about 250 per 100,000 people
C about 500 per 100,000 people

4 What does this graph suggest for the number of deaths from flu or pneumonia in the year 2050?

F 200 per 100,000 people
G 100 per 100,000 people
H 30 per 100,000 people

5 Which of these is most likely to be the leading cause of death in the year 2000?

A heart disease
B flu/pneumonia
C tuberculosis

6 Which of these diseases will probably be a cause of death for fewer Americans in the year 2010 than today?

F heart disease
G flu/pneumonia
H cancer

7 Which of the following generalizations is supported by the information in this graph?

A Factory waste is causing a rise in cancer deaths.
B People are not eating as healthily as they used to.
C Deaths from communicable diseases (like flu and tuberculosis) are falling.

Data Interpretation Skills Practice

This graph shows how the percentage of households with computers and cable has changed since 1984. Study the graph. Then do Numbers 1 through 6.

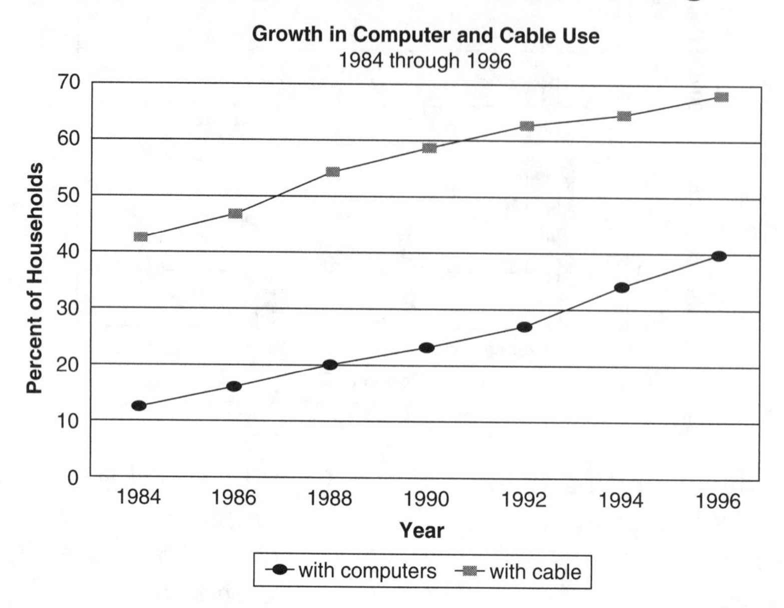

1 Which of these is the best estimate of how many more households had cable in 1996 than had it in 1984?

A 15 percent more
B 19 percent more
C 25 percent more
D 80 percent more

2 In 1996, there were about 100 million households in the U.S. How many of them had cable?

F about 68 million
G about 6.8 million
H about 680
J There is no way to tell.

3 In which year was there the greatest difference between the number of households with cable and those with computers?

A 1984
B 1990
C 1994
D 1996

4 During which of these periods was there the least increase in the number of households with computers?

F 1984 to 1986
G 1986 to 1988
H 1988 to 1990
J 1990 to 1992

5 Which of the following is the best estimate of the *median* percentage of households that used cable during the years shown?

A 47%
B 50%
C 63%
D 58%

6 Use the trends shown in this graph to predict a value for the percent of households with computers in 1998.

F 42%
G 47%
H 53%
J 38%

The United States controls several islands outside its own borders. This table shows the size and population of each island territory. Study the table. Then do Numbers 7 through 11.

U.S. Territories

Territory	1990 Population	1996 Population	Land Area (in square miles)
American Samoa	47,000	60,000	77
Guam	133,000	157,000	210
Northern Mariana Islands	43,000	52,000	179
Puerto Rico	3,522,000	3,783,000	3,427
Virgin Islands	102,000	97,000	134

Source: U.S. Census Bureau

7 What is the largest U.S. territory outside the states?

A American Samoa
B Virgin Islands
C Guam
D Puerto Rico

8 Which U.S. territory lost population between 1990 and 1996?

F American Samoa
G Northern Mariana Islands
H Puerto Rico
J Virgin Islands

9 How many more people lived in Guam in 1996 than in 1990?

A 124,000
B 157,000
C 24,000
D This cannot be determined from the information in the table.

10 Population density is *the number of people per square mile.* Which of the following number sentences could you use to find the population density of the Northern Mariana Islands in 1996?

F $179 \times 52{,}000 =$ ______
G $179 \div 52{,}000 =$ ______
H $52{,}000 - 43{,}000 =$ ______
J $52{,}000 \div 179 =$ ______

11 Which territory has about $\frac{1}{3}$ as much land as Guam?

A American Samoa
B Northern Mariana Islands
C Puerto Rico
D Virgin Islands

Algebra

Patterns

A key to understanding math is to see and use patterns. When you describe a pattern using words and symbols, you are using the math ideas called **algebra.**

PRACTICE

Below, the shaded regions form a pattern. Find the pattern in each box. Then follow that pattern to shade the figure at the right in each box.

1 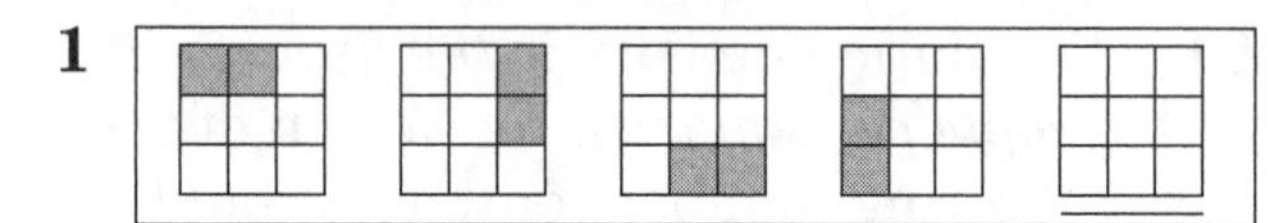

2 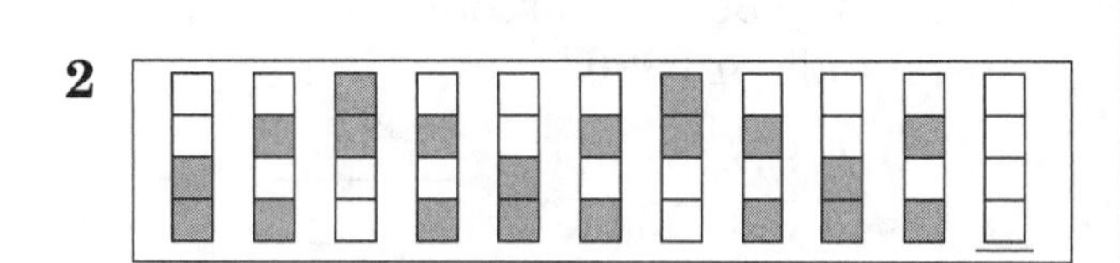

3 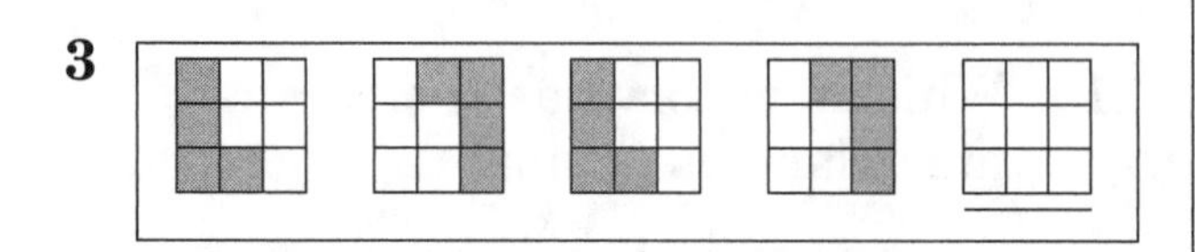

4 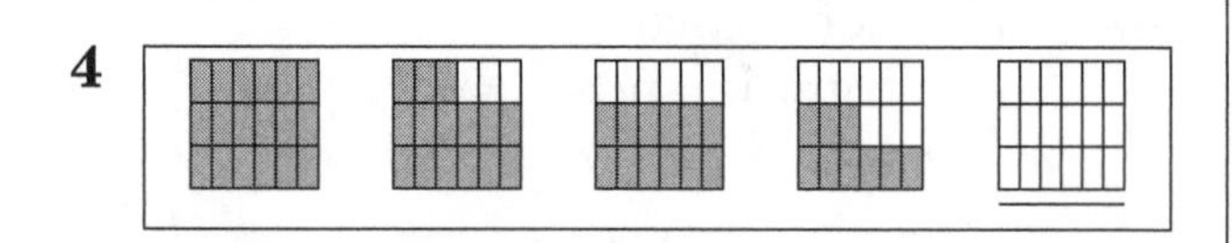

5 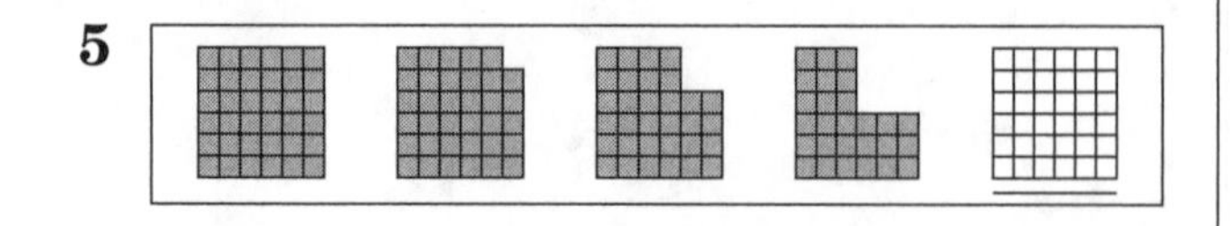

Find the pattern in each box below. Then draw the next figure in the pattern.

6 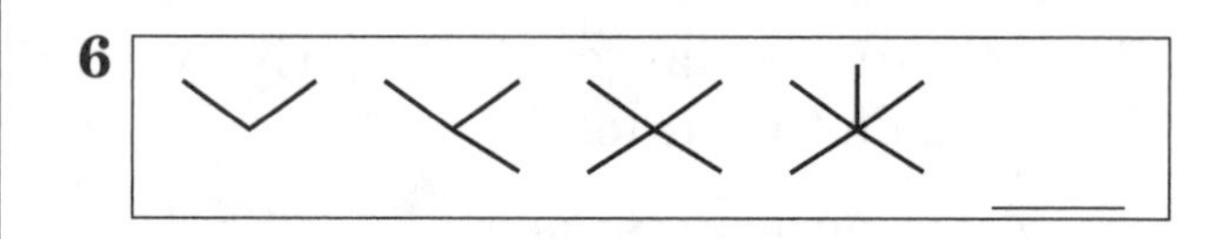

7 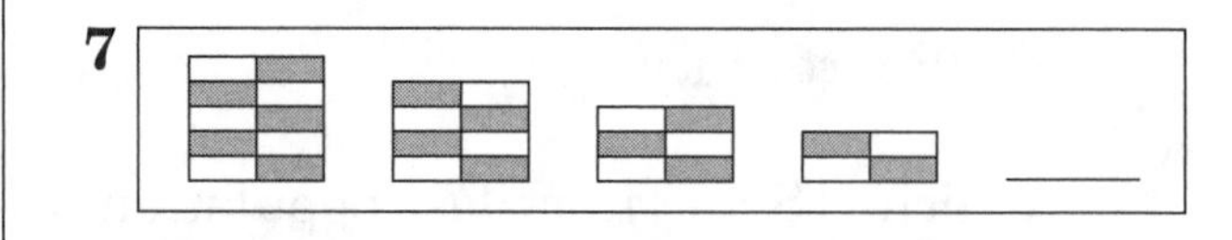

8 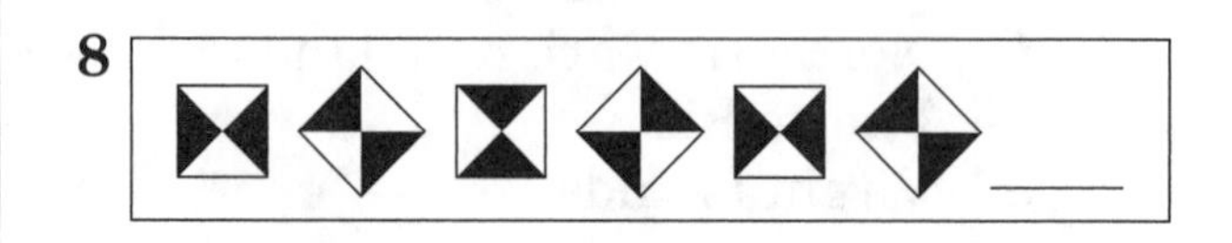

9 

10 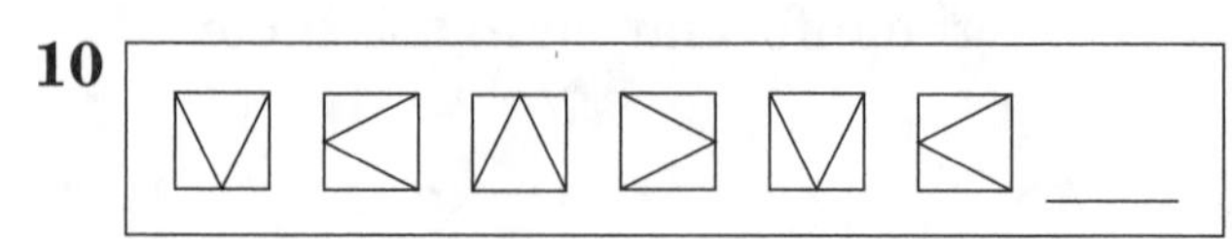

Finding Number Patterns

You can create a pattern with numbers as well as with shapes. Here are some number pattern problems.

PRACTICE

For each problem, fill in the blanks.

1 164, 160, 156, 152, 148, . . .
This pattern starts with 164 and subtracts 4 over and over again. Are the numbers in this list odd or even? ______

2 2, 6, 18, 54, 162, . . .
This pattern starts with 2 and multiplies each number by 3. Every number in this list can be divided by _?_. ______

3 3, 9, 15, 21, 27, . . .
This pattern starts with 3 and adds 6 to each number. Every number in this list can be divided by _?_. ______

4 105, 95, 85, 75, 65, 55, . . .
This pattern starts with 105 and keeps subtracting 10. Circle the number that will be somewhere in the list.
10 25 33 42 ______

5 Start with 5 and add 3 to each number:
5, 8, 11, . . .
What are the next three numbers ______

6 Start with 5. Multiply by 2 and then subtract 3:
5, 7, 11, . . .
What are the next three numbers? ______

7 Here are some multiples of 9:
9, 18, 27, . . .
Every multiple of 9 can be divided by 9 and _?_. ______

8 24, 18, 12, 6, 0, . . .
In this pattern, each new number is the last number _?_. (Give an operation and a number.) ______

9 400, 200, 100, 50, 25, . . .
In this pattern, each number is the last number _?_. ______

10 130, 115, 100, 85, 70, 55, 40, . . .
In this pattern, each number is the last number _?_. ______

11 36, 41, 46, 51, 56, 61, . . .
In this pattern, each number is the last number _?_. ______

12 2, 5, 8, 11, 14, 17, . . .
In this pattern, each number is the last number _?_. ______

13 5, 17, 29, 41, 53, 65, . . .
In this pattern, each number is the last number _?_. ______

14 There are even numbers:
2, 4, 6, 8, . . .
Every even number can be divided by _?_. ______

Patterns in Number Sentences

The number sentences below are incomplete. Use what you know about number facts to answer each question.

PRACTICE

Write +, –, ×, or ÷.

Example 1

The sentence 2 ☐ 3 = 5 needs the symbol "+" because 2 [+] 3 = 5.

1. 9 ☐ 2 = 11
2. 20 ☐ 3 = 17
3. 18 ☐ 5 = 23
4. 10 ☐ 10 = 0
5. 9 ☐ 2 = 18
6. 30 ☐ 5 = 25
7. 50 ☐ 25 = 25
8. 5 ☐ 5 = 25
9. 95 ☐ 25 = 70
10. 39 ☐ 13 = 3
11. 4 ☐ 9 = 36
12. 100 ☐ 10 = 10
13. 6 ☐ 7 = 42
14. 35 ☐ 5 = 3 + 4
15. 2 ☐ 6 = 12 – 0

Write a number for each sentence.

Example 2

For 5 + ☐ = 9, write 4 because 5 + [4] = 9.

16. 3 × 3 × ☐ = 27
17. 13 + ☐ = 15
18. 8 + ☐ = 13
19. 5 + ☐ = 25
20. 5 + 3 + ☐ = 16
21. 50 – ☐ = 20
22. 16 – ☐ = 6
23. ☐ – 6 = 4 ← *Think:* What is 4 + 6?
24. ☐ – 10 = 5
25. ☐ – 12 = 20
26. 20 – ☐ = 8
27. 14 × ☐ = 28 ← *Think:* What is 28 ÷ 14
28. 3 × ☐ = 33
29. 2 × 3 × ☐ = 60 ← *Think:* What is 60 ÷ 6?

Some Basic Number Properties

<table>
<tr>
<td>The order of the numbers in a problem does not matter in addition and multiplication problems but does matter in subtraction and division problems.

5 + 7 and 7 + 5 have the same value.
5 × 7 and 7 × 5 have the same value.
7 − 5 and 5 − 7 have different values.
15 ÷ 5 and 5 ÷ 15 have different values.</td>
<td>Addition and subtraction are inverse operations, which means that each one will undo the other. Similarly, multiplication and division are inverse operations.

+ 2 + 3 = 5, so 5 − 3 = 2
− 7 − 3 = 4, so 4 + 3 = 7
× 6 × 7 = 42, so 42 ÷ 7 = 6
÷ 50 ÷ 10 = 5, so 5 × 10 = 50</td>
</tr>
<tr>
<td colspan="2">There are two properties of one.</td>
</tr>
<tr>
<td>First, if you multiply or divide a number by one, you do not change the value.
15 × 1 = 15 20 ÷ 1 = 20</td>
<td>Second, if you divide any (nonzero) number by itself, the quotient is one.
15 ÷ 15 = 1</td>
</tr>
<tr>
<td colspan="2">There are several properties of zero.</td>
</tr>
<tr>
<td>If you add or subtract zero from a number, you do not change the value.
15 + 0 = 15 20 − 0 = 20

If you subtract a number from itself, the difference is zero. 15 − 15 = 0</td>
<td>If you multiply any number by zero, the product is zero. 20 × 0 = 0

If you divide zero by any (nonzero) number, the quotient is zero. 0 ÷ 15 = 0

An expression like “15 ÷ 0” has no meaning.</td>
</tr>
</table>

PRACTICE

Write +, −, ×, ÷, or a number for each number sentence.

1 514 ☐ 514 = 1

2 74 ☐ 74 = 0

3 0 ☐ 21 = 0

4 112 ☐ 0 = 112

5 817 ☐ 0 = 0

6 915 ÷ 1 = ☐

Below, write +, −, ×, or ÷ and a number.

7 817 + 35 = 852, so 852 ☐ = 817

8 718 − 26 = 692, so 718 ☐ = 26

Below, the same number goes into each box. Tell whether the statement is *true* or *false.*

9 If 89 − ☐ = 45, then 45 + ☐ = 89.

10 If ☐ + 38 = 94, then 94 − ☐ = 38.

Functions

Here is a simple rule: Add 4. If you apply that rule to the numbers 0, 5, and 10, you get 4, 9, and 14, respectively. A rule that changes one value to another value is called a **function.**

PRACTICE

For each problem, tell the rule that works for all 3 boxes.

Example Add 4.

3 ☐ = 7
1 ☐ = 5
5 ☐ = 9

1 5 ☐ = 11 ______
1 ☐ = 7
3 ☐ = 9

2 35 ☐ = 10 ______
25 ☐ = 0
33 ☐ = 8

3 7 ☐ = 21 ______
3 ☐ = 9
5 ☐ = 15

For Numbers 4–6, the rule tells how to change each "In" number to its corresponding "Out" number. Use the rule to find the missing number.

Example: Subtract 2 and then multiply by 3. The missing number is 12.

Input	5	2	4	3	6
Output	9	0	6	3	

4 Multiply by 2 and then add 1.

Input	1	2	5	4
Output	3	5	11	

5 Divide by 2 and then add 5.

Input	12	8	6	10
Output	11	9	8	

6 Subtract 3 and then multiply by 6.

Input	4	10	8	6
Output	6	42	30	

Circle the letter for the phrase that describes each function below.

7

Input	1	2	3	4
Output	6	9	12	15

F Add 1 and then multiply by 3.
G Multiply by 5 and then add 1.
H Add 5.

8

Input	1	2	3	4
Output	$\frac{1}{2}$	1	$1\frac{1}{2}$	2

A Subtract $\frac{1}{2}$.
B Multiply by 2 and then subtract $1\frac{1}{2}$.
C Divide by 2.

Writing Letters and Symbols for Words

An algebra problem uses letters to take the place of numbers. These letters are called **variables** or **unknowns.** A letter can represent a single number or it can represent many values.

In	3	4	5	7
Out	1	2	3	5

In this table, suppose that x represents each "In" number. Then each "Out" number can be described as $x - 2$.

Example:

a number increased by five

If you use the letter *n* to represent "a number," then this phrase can be described as $n + 5$.

Follow these steps to write an expression using the letters and symbols of algebra:
1. Decide whether the words describe something being added, subtracted, multiplied or divided.
2. Identify the numbers involved in the problem.
3. Write the numbers and the proper signs (+, –, ÷, or ×). Remember, the order of numbers is important in a subtraction problem or a division problem.

PRACTICE

Write an algebra expression for each phrase. Let the letter x stand for the unknown number. (*Note:* In algebra, the expression $3 \times x$ is usually written without the multiplication sign as $3x$. For the problems below, show multiplication without using the × symbol.)

1 a number multiplied by sixteen ______

2 a number increased by twelve ______

3 five times a number ______

4 a number plus seventeen ______

5 six less than a number ______

6 the sum of a number and ten ______

7 half of a number (or a number divided by 2) ______

8 the difference between a number and twenty ______

9 one-eighth of a number ______

10 a number divided by three (or $\frac{1}{3}$ of a number) ______

11 five more than a number ______

12 two less than a number ______

13 twice a number ______

14 five times a number ______

15 a number plus six ______

16 five divided by a number ______

Writing Two-Step Algebraic Expressions

Algebra is especially useful for solving complicated word problems. A long word problem can be summed up with just a few algebra symbols.

Problem: A number is multiplied by twelve, and then three is taken away.
Algebraic Expression:
$12n - 3$

Problem: When you divide a number by 4 and then add 16, the result is 13. What is the number?
Algebraic Equation:
$\frac{n}{4} + 16 = 13$

Problem: Angela is shopping. She begins with \$30.00 in her purse, and she must save \$1.50 to pay for her train ride home. How much money does she have left to spend after paying \$5.50 to see a movie?
Algebraic Equation:
$\$30.00 - \$5.50 - x = \$1.50$

Guidelines for Writing Complex Algebraic Expressions

Multiplication of 3 and x is usually shown with a raised dot ($3 \cdot x$) or with no sign at all ($3x$).

Division of 5 by 3 is usually shown in fraction form as $\frac{5}{3}$.

If it is not clear which step in a problem should be done first, the first step is enclosed in parentheses:
$(9 - 2) + 5$
That expression means "Subtract two from nine, and *then* add five to the difference."

To show a fraction of a variable, write the variable in the numerator of the fraction:

$\frac{1}{3}$ of $n = \frac{1 \cdot n}{3} = \frac{n}{3}$

$\frac{2}{3}$ of $n = \frac{2 \cdot n}{3} = \frac{2n}{3}$

PRACTICE

Circle the correct algebraic expression for each phrase below.

1 Multiply two times a number and then subtract twelve.

$2n - 12$ $\quad 2(n - 12)$ $\quad 2n - 12n$

2 Divide a number by four and then add six.

$\frac{n}{4} + 6$ $\quad 4n + 6$ $\quad \frac{4n}{6}$

Write your own algebraic expression for each phrase below. Use *n* for the unknown number.

3 Add fifteen to two times a number. ______

4 Add two to one-third of a number. ______

5 Add 5 to a number and then subtract 1 from the sum. ______

Using Algebra to Solve Word Problems

Write an algebraic expression for each situation below. *Hint:* Look for the key words that signal what sign you should use. There is a complete list of signal words on page 23.

1 Linda spends d dollars. Then she spends 15 dollars more. How many dollars has Linda spent in all?

____

2 Warren is driving to Phoenix with three other people. It will take them h hours. If they divide the driving time up evenly, how many hours will Warren drive?

____

3 Laurie Ann works 8 hours a day. So far today she has worked x hours. How many hours are left in her workday?

____

4 Ken's cat eats c cups of food a day. Ken is leaving town for 6 days. How many cups of food will the cat eat while Ken is gone?

____

5 Aisha is 6 inches taller than her husband. Aisha is t inches tall. How many inches tall is her husband?

____

6 Whitney is speaking to four different classes at the local grade school. She will spend m minutes with each class, and she will have to wait 30 minutes before seeing the last group. How many minutes will Whitney spend at the school?

____

To solve a word problem using algebra, you must write an **equation.** An equation is an expression such as $\mathbf{3n - 5 = 10}$. The equal sign in the equation says the same value is on each side of that sign.

For each problem below, circle the letter for the correct explanation.

7 Enrique earns $24,000 per year. In the equation $24{,}000 \div 12 = x$, what is x?

A the amount he earns per hour
B the amount he earns in 12 years
C the amount he earns per month

8 Regular unleaded gasoline costs $1.15 per gallon. In the equation $\$1.15 \times 10 = x$, what is x?

F the cost of 10 gallons of gas
G the amount of gas you can buy for 10 dollars
H the amount of gas you need to drive 10 miles

9 Rodney has $150 to spend on Christmas presents this year. He puts $50.00 aside to buy a gift for his wife. He divides the remaining money evenly to buy gifts for his three children. In the equation $\$150 - \$50.00 = x$, what is x?

A the number of gifts he will buy
B the amount of money he will spend on each child
C the amount of money he will spend on all the children together

Solving Equations

The equation $\mathbf{x + 3 = 15}$ means "add an unknown number and 3; the result is 15." To solve that equation, perform the *inverse* (the opposite) of the operation shown. The inverse of "add 3" is "subtract 3." To solve the equation, subtract 3 from each side of the equation.

To solve an addition problem, you subtract.

Equation	$x + 3$	$=$	15
Subtract 3 from each side.	-3		-3
	$x + 0$		12
Adding 0 to a number does not change its value, so	x	$=$	12

To solve a subtraction problem, you add.

$$\begin{array}{rcr} x - 3 & = & 6 \\ +3 & & +3 \\ \hline x & = & 9 \end{array}$$

Some problems need two steps.

$$\begin{array}{rcr} 15 - x & = & 7 \\ +x & & +x \\ \hline 15 & = & 7 + x \\ -7 & & -7 \\ \hline 8 & = & x \end{array} \quad \text{or } x = 8$$

> ***Remember:***
> **To solve an equation, you must get the variable alone on one side of the equation.**

PRACTICE

Solve each of the following equations. Then check your answer by seeing whether it makes the original equation true. *Remember:* Always do the same operation to both sides of an equation.

1 $x + 9 = 27$

2 $x + 13 = 65$

3 $x + 19 = 25$

4 $x - 36 = 52$

5 $x - 16 = 45$

6 $x - 51 = 24$

7 $12 + x = 37$

8 $35 - x = 19$

9 $44 - x = 24$

To solve a multiplication equation, you can divide.	To solve a division equation, you can multiply.	To check your answer, replace the variable with your solution. Then work *separately* on each side of the equation.
Problem: $2x = 24$ Divide by 2: $\frac{2x}{2} = \frac{24}{2}$ $x = 12$	**Problem:** $\frac{x}{4} = 6$ Multiply by 4: $(4)(\frac{x}{4}) = (4)(6)$ $x = 24$	To check $x = 24$, in the equation $\frac{x}{4} = 6$, substitute 24 for x. Does $\frac{24}{4} = 6$? Does $6 = 6$? Yes. The solution $x = 24$ checks.

Solve each equation below. Then check your work.

10 $4n = 64$

11 $75 = 5x$

12 $6a = 72$

13 $n \div 6 = 6$

14 $b \times 8 = 96$

15 $28 \div a = 14$

16 $\frac{x}{6} = 11$

17 $\frac{x}{10} = 5$

18 $24 + g = 49$

These equations have two steps each. To solve them, first reverse the addition or subtraction. Then reverse the multiplication or division.

Equation: $2x + 5 = 15$

Subtract 5: $-5 \quad -5$

$2x = 10$

Divide by 2: $\frac{2x}{2} = \frac{10}{2}$

Solution: $x = 5$

19 $3x - 2 = 7$

20 $4n + 12 = 92$

21 $\frac{b}{5} + 6 = 36$

22 $\frac{t}{2} - 8 = 2$

Write an equation for each problem below. Then solve and check your work.

23 A number times 3 is 15. What is the number?

24 One-half of a number is 12. What is the number?

25 A number plus 10 is 92. What is the number?

26 Five more than three times a number is 47. What is the number?

27 If you take one-third of a number and then add 2, the result is 7. What is the number?

Algebra Skills Practice

Circle the letter for the correct answer to each problem.

1 What number goes in the box to make the number sentence true?

$$4 + 5 + \square = 12$$

A 7
B 8
C 3
D 4

2 Tomato soup costs 33 cents per can. Which of these number sentences shows how much 4 cans would cost?

F 33 cents $\times$ 4 = $\square$
G 33 cents $\div$ 4 = $\square$
H 4 $\div$ 33 cents = $\square$
J 33 cents + 4 = $\square$

3 Find the rule that will change each "Input" number to the corresponding "Output" number. Then use the rule to find the missing number.

Input	40	26	44	20
Output	20	13	22	

A 10
B 40
C 0
D 15

4 Tyrone has \$5.00, and Coney dogs cost \$0.95 each. If Tyrone solves the equation \$5.00 $\div$ \$0.95 = x, what will the value of x represent?

F the cost of 0.95 Coney dogs
G the money he would have left after buying a Coney dog
H the number of Coney dogs he can buy for \$5.00
J the price of 5 Coney dogs

5 If you start with 3, then add 3 to that number, and repeat adding 3 to each new number, which number will *never* be in the pattern?

A 9
B 15
C 18
D 22

6 For which of these equations is x equal to 8?

F $8 - 1 = x$
G $8 \div 1 = x$
H $8 \times 0 = x$
J $8 + 1 = x$

7 Elliot has 15 boxes of candy. He gives 3 to his mother and splits the remaining boxes with his best friend. Which of these equations shows how to figure out how many boxes Elliot has left?

A $\frac{15}{2} - 3 = x$

B $\frac{15}{2} + 3 = x$

C $15 - \frac{3}{2} = x$

D $\frac{15 - 3}{2} = x$

8 City workers picked up yard waste on April 6, April 15, April 24, and May 3. If this pattern continues, when will be the next date they pick up yard waste?

F May 6
G May 12
H May 14
J May 17

9 Last week Gary worked 55 hours. He earned $14 per hour, plus he got a $100 bonus. Which of these number sentences shows how to find out how much Gary earned altogether?

A (55 × $14) + $100 = _______
B 55 × ($14 + $100) = _______
C 55 + $14 + $100 = _______
D (55 × $14) + (55 × $100) = _______

10 What number goes in the box to make the number sentence true?

$7 \times \square = 0$

F 7
G 1
H $\frac{1}{7}$
J 0

11 What number is missing from the table below?

Rule: Divide by 2 and then add 1.

Input	10	12	16	18
Output	6	7	9	

A 9
B 12
C 10
D 19

12 In which of these equations is x equal to 12?

F $x \div 2 = 24$
G $x - 5 = 17$
H $x \times 3 = 4$
J $x + 6 = 18$

Use the information below to do Numbers 13 and 14.

There are 3 women on Tanika's work crew. Yesterday they worked 6 hours and they assembled 500 CD players.

13 Tanika used this equation to figure out how fast the women worked yesterday.

$500 \div 6 = n$

What does the n represent in the equation?

A the number of CD players they made per worker
B the number of CD players they made per hour
C the number of minutes they spent on each CD player
D the number of CD players they made each minute

14 Each woman on the crew earns $8.50 per hour. Which of the following equations could you use to figure out the crew's total earnings yesterday?

F $8.50 × 3 = _______
G $8.50 × 6 = _______
H $8.50 × 3 + 6 = _______
J $8.50 × 3 × 6 = _______

Measurement

Reading a Scale

A number line on a measurement tool is called a **scale.** To use a simple scale, such as a ruler, follow these steps:

1. Line up the zero with one end of the object you are measuring.
2. Find the number that lines up with the other end of the object.
3. Check your work by measuring the object again.

Sometimes a scale skips numbers. In that situation, you can estimate the measurement shown on the scale.

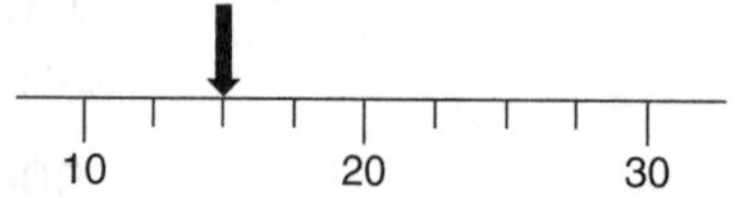

This pointer is about halfway between 10 and 20, so the measurement shown on the scale is about 15.

PRACTICE

Find each measurement below.

1 The pointer on a scale is about halfway between 18 and 20. What number is halfway between 18 and 20? _______

2 The pointer on a scale is about halfway between 90 and 100. What number is halfway between 90 and 100? _______

3 The pointer on a scale is about halfway between 60 and 80. What number is halfway between 60 and 80? _______

4 The pointer on a scale is about $\frac{1}{4}$ of the way from 60 to 80. What measurement is shown? _______

5 The pointer on a scale is about $\frac{1}{4}$ of the way from 0 to 100. What measurement is shown? _______

6 The pointer on a scale is about $\frac{1}{3}$ of the way from 60 to 90. What measurement is shown? _______

7 What measurement is shown on this scale? _______

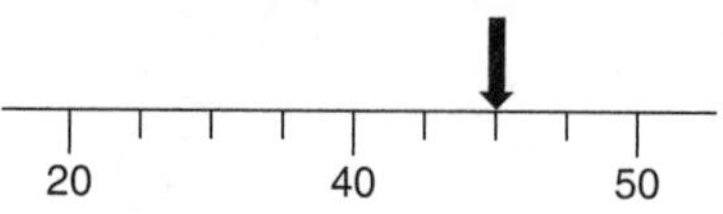

8 What measurement is shown on this scale? _______

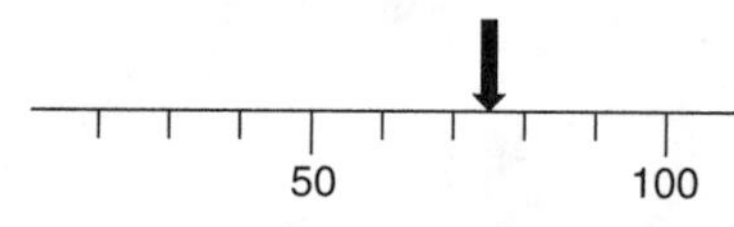

If the scale has tick marks, count the number of spaces formed between numbers. Then figure out what each tick mark is worth.

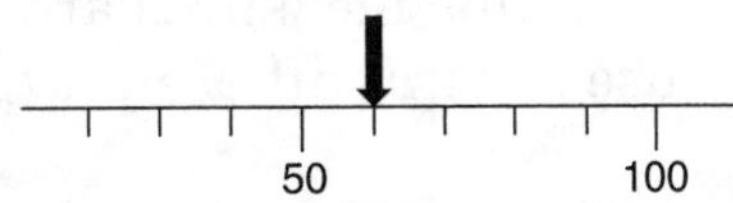

In the scale below, there are 5 tick marks between 50 and 100, so each tick mark represents 10 units.

The arrow is 1 tick mark beyond 50, so the reading is 50 + 10, or 60 units.

PRACTICE

9 On a scale, the interval between 30 and 40 is divided into 5 parts. How many units does each tick mark represent? ______

10 On a scale, the interval between 60 and 80 is divided into 10 parts. How many units does each tick mark represent? ______

11 On a scale, the interval between 75 and 100 is divided into 5 parts. How many units does each tick mark represent? ______

12 On a scale, the interval between 50 and 100 is divided into 5 parts. Each tick mark represents _?_ units. ______

13 On a scale, the interval between 0 and 100 is divided into 5 parts. Each tick mark represents _?_ units. ______

14 On a scale, the interval between 20 and 40 is divided into 5 spaces. Each tick mark represents _?_ units. ______

15 On a scale, the interval between 0 and 100 is divided into 4 spaces. Each tick mark represents _?_ units. ______

16 Each tick mark below represents _?_ units, and the reading on the scale is _?_. ______

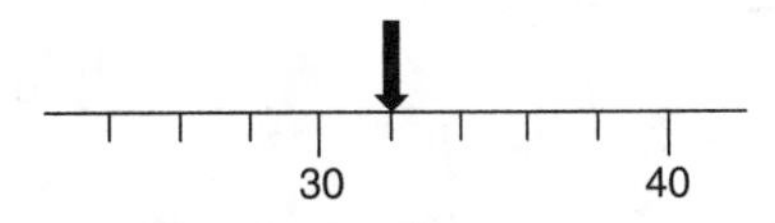

17 Each tick mark below represents _?_ units, and the reading on the scale is _?_. ______

Using a Ruler

Most standard rulers, yardsticks, and tape measures are not as simple as the scales you used on the previous pages. They use several different types of tick marks:

You will see the following on most rulers marked in inches and feet:

- The longest tick marks divide the ruler into inches.
- The next longest tick marks divide each inch into half inches.
- The half inches are divided into fourths by slightly shorter tick marks.
- The fourths are divided into eighths by even shorter tick marks.
- The shortest tick marks may be eighths of an inch or sixteenths of an inch.

PRACTICE

Find a ruler marked in inches. Look at the long tick mark halfway between each pair of numbers. Use those long tick marks to measure each line below to the nearest $\frac{1}{2}$ inch. Some answers will be whole numbers.

1 ————————

_______ inches

2 ———————————

_______ inches

3 ———

_______ inch

4 ——————————————

_______ inches

5 —————

_______ inch

Find the tick marks on your ruler that divide each inch into four equal parts, or fourths. Use those marks to measure the lines below to the nearest $\frac{1}{4}$ inch.

6 —

_______ inch

7 ————

_______ inch

8 ———————

_______ inches

9 ———

_______ inch

10 ——————————

_______ inches

Find a ruler marked in inches. Look at the long tick mark halfway between each pair of numbers. Use those long tick marks to measure each line below to the nearest $\frac{1}{2}$ inch. Some answers may be whole numbers.

11 ———

_______ inch

12 ——————

_______ inch

13 ——

_______ inch

14 —

_______ inch

15 ————

_______ inch

Circle the letter for the most accurate measurement of each line below.

16 —

A $\frac{1}{2}$ inch **C** $\frac{1}{16}$ inch

B $\frac{5}{16}$ inch **D** $\frac{3}{16}$ inch

17 ——

F $\frac{3}{4}$ inch **H** $\frac{1}{4}$ inch

G $\frac{7}{16}$ inch **J** $\frac{5}{8}$ inch

Express each of the measurements below in lowest terms. If you need to review reducing or adding fractions, use pages 44, 46, 48, 51, and 52.

Example

$\frac{12}{8}$ in. = $1\frac{1}{2}$ in.

18 $\frac{8}{16}$ in. = _______ in.

19 $\frac{6}{8}$ in. = _______ in.

20 $\frac{6}{4}$ in. = _______ in.

21 $\frac{14}{16}$ in. = _______ in.

22 $\frac{4}{16}$ in. = _______ in.

23 $\frac{6}{2}$ in. = _______ in.

24 $\frac{7}{8}$ in. + $\frac{1}{16}$ in. = _______ in.

25 $\frac{3}{8}$ in. + $\frac{1}{16}$ in. = _______ in.

26 $\frac{5}{8}$ in. + $\frac{1}{16}$ in. = _______ in.

27 $\frac{1}{2}$ in. + $\frac{1}{16}$ in. = _______ in.

28 $\frac{3}{4}$ in. + $\frac{1}{16}$ in. = _______ in.

Measuring Real Objects

The problems below ask you to find objects in your home or your classroom and measure them with your own standard ruler.

1 Find the height and width of *two* of the objects listed below. Give your answers to the nearest $\frac{1}{4}$-inch.

a cassette tape ______ by ______

a new #2 pencil ______ by ______

a standard index card ______ by ______

a regular brown paper grocery bag (flattened) ______ by ______

2 The objects below are difficult to measure because they have rounded shapes. Measure the *widest and tallest* parts of two of the objects listed. Give your answers to the nearest $\frac{1}{2}$-inch.

a 12-ounce aluminum soda can ______ by ______

a 1-gallon plastic milk jug ______ by ______

a 12-ounce can of frozen juice ______ by ______

a plastic 2-liter soda bottle ______ by ______

3 Measure the height and width of *two* objects listed below. Give your answers to the nearest $\frac{1}{8}$-inch.

a can of Campbell's® condensed soup ______ by ______

a postage stamp ______ by ______

a computer disk ______ by ______

a dollar bill ______ by ______

a 9-volt battery ______ by ______

Circle the letter for the most accurate measurement of each object.

4 the width of a dime

A $\frac{11}{16}$ inch **C** $\frac{1}{2}$ inch

B $\frac{3}{4}$ inch **D** $\frac{7}{8}$ inch

5 the width of a penny

F $\frac{4}{8}$ inch **H** $\frac{3}{4}$ inch

G $\frac{7}{8}$ inch **J** $\frac{14}{16}$ inch

6 the width of a nickel

A 1 inch **C** $\frac{3}{4}$ inch

B $\frac{5}{8}$ inch **D** $\frac{13}{16}$ inch

Standard Units of Measure

One system of measurement uses **standard units.**

Standard Units of Measurement

Temperature degrees Fahrenheit (°F)	Normal body temperature is 98.6°F. Water boils at 212°F. Water freezes at 32°F.
Length 12 inches (in.) = 1 foot (ft) 3 feet = 1 yard (yd) 5,280 feet = 1 mile (mi)	An inch is about the length of a straight pin. A foot is about the length of a man's foot. A yard is about the length of your arm. A mile is 8–10 city blocks.
Weight 1 pound (lb) = 16 ounces (oz) 1 ton (T) = 2,000 lb	A pencil weighs about 1 ounce. An eggplant weighs about 1 pound. A car weighs about 1 ton.
Capacity 1 pint = 2 cups 1 quart = 4 cups (or 2 pints) 1 gallon = 4 quarts (or 16 cups)	An ice cream dish holds 1 cup. A mug holds about 1 pint. A thin milk carton holds 1 quart. A plastic milk jug holds 1 gallon.

Time
1 minute (min) = 60 seconds (sec)
1 hour (hr) = 60 min
1 day = 24 hr

PRACTICE

Circle the letter for the best estimate for each measurement.

1 the temperature in a refrigerator

A 70°F
B 25°F
C 40°F

2 the width of a dime

F 3 inches
G $\frac{1}{2}$ inch
H 6 inches

3 the weight of a dinner roll

F one pound
G one ounce
H $\frac{1}{2}$ pound

4 the capacity of a soup can

A $1\frac{1}{2}$ cups
B 1 gallon
C 1 quart

Circle the *greater* measurement in each pair.

5	15 inches	1 foot
6	5 feet	2 yards
7	12 ounces	1 pound
8	22 ounces	1 pound
9	3 cups	1 pint
10	3 cups	1 quart
11	1 gallon	12 cups

Converting Within the Standard System

One way to convert between units is to use a proportion. Put like units in the numerator and like units in the denominator.

To review proportions, see pages 69–71.

Problem: How many yards are in 15 feet?
Solution:

Set up a proportion. $\frac{1 \text{ yd}}{3 \text{ ft}} = \frac{? \text{ yd}}{15 \text{ ft}}$

Cross multiply. $? \times 3 = 1 \times 15$

Divide both sides by the number 3. $\frac{? \times 3}{3} = \frac{1 \times 15}{3}$

$? = 5$

Another way to convert between units is to use multiplication or division. Look back at the table on page 111.

To change a measurement to smaller units, you can multiply.	To change a measurement to larger units, you can divide.
Examples:	**Examples:**
To change feet to inches, multiply by 12.	To change feet to yards, divide by 3.
To change pounds to ounces, multiply by 16.	To change pounds to tons, divide by 2,000.

PRACTICE

Convert each measurement below to the given unit.

1 9 ft = _______ yd

2 24 in. = _______ ft

3 36 in. = _______ yd

4 $\frac{1}{2}$ ft = _______ in.

5 $\frac{1}{3}$ yd = _______ ft

6 $1\frac{1}{2}$ feet = _______ in.

7 $1\frac{2}{3}$ yd = _______ ft

8 1 ft 2 in. = _______ in. (Convert 1 foot into inches and then add 2.)

9 2 yd 2 ft = _______ ft (Convert the yards into feet and then add 2.)

10 2 ft 6 in. = _______ in.

11 3 ft 5 in. = _______ in.

12 11 ft = ___ yd ___ ft (Divide 11 by 3 and give the remainder in feet.)

13 26 in. = ___ ft ___ in.

14 3 lb = _______ oz

15 $\frac{1}{2}$ lb = _______ oz

16 23 oz = ___ lb ___ oz

17 2 lb 3 oz = _______ oz

18 360 min = _______ hr

19 $1\frac{1}{2}$ hr = _______ min

20 $\frac{1}{4}$ hour = ___ minutes

The Metric System

The **metric system of measurement** uses **metric units.** Most countries around the world use the metric system.

Metric Units of Measurement

Temperature Degrees Celsius (°C)	Water boils at 100°C. Water freezes at 0°C. Normal body temperature is 37°C.
Length 1 meter (m) = 1,000 millimeters (mm) = 100 centimeters (cm) 1 cm = 10 mm 1 kilometer (km) = 1,000 m	A needle is about 1 millimeter wide. A kindergarten student is about 1 meter tall. Your little finger is about 1 centimeter wide.
Weight 1 gram (g) = 1,000 milligrams (mg) 1 kilogram (kg) = 1,000 g	A needle weighs about 1 milligram. A peanut weighs about 1 gram. A city telephone book weighs about 1 kg.
Capacity 1 liter (L) = 1,000 milliliters (mL) 1 kiloliter (kL) = 1,000 L	A large plastic soda bottle holds 2 liters. A dose of cough medicine is about 10 mL. A septic tank holds about 2 kiloliters.

The metric system is based on **powers of ten.** (The powers of ten are 1, 10, 100, 1,000, 10,000, and so on.) The basic units are **meter, gram,** and **liter.** Any measurement that begins with ***milli-*** is one-thousandth of the basic unit. Any measurement that begins with ***kilo-*** is one thousand times the basic unit.

PRACTICE

Fill in each blank.

1 A pot of coffee weighs about 1 _______.

2 A coffee mug holds about $\frac{1}{4}$ _______.

3 It takes about 5 ____ of oil to fry an egg.

4 A golf tee is about 5 _______ long.

5 A baseball bat is about 1 _______ long.

6 A cat weighs about 4 _______.

7 $\frac{1}{2}$ kg = _______ g

8 2 L = _______ mL

9 3,000 mL = ______ L

10 $1\frac{1}{2}$ m = _______ mm

11 3,500 mm = ______ m

12 10 cm = _______ mm

13 $2\frac{1}{2}$ kL = _______ L

14 150 cm = _______ m

15 3 m 15 cm = ____ cm

16 1 kg 450 g = _____ g

17 3 L 15 mL = ___ mL

18 1,515 g = __ kg __ g

Comparing Standard and Metric Units

You should have a general idea of how standard and metric units compare. The symbol "≈" means "is approximately equal to."

1 inch ≈ 2.5 cm 1 meter ≈ 1.1 yards 1 kilometer ≈ 0.6 mile	1 kilogram ≈ 2.2 lb 1 ounce ≈ 28 g	1 liter ≈ 1.1 qt 1 kiloliter ≈ 275 gal

On the previous pages you converted measurements *within* the standard system and *within* the metric system. You can use the same process to get equivalent measurements *between* the two systems.

Problem: 40 cm ≈ _?_ inches

Solution:

Set up a proportion. $\frac{1 \text{ in.}}{2.5 \text{ cm}} = \frac{? \text{ in.}}{40 \text{ cm}}$

Cross multiply. $2.5 \times ? = 1 \times 40$

Divide by 2.5. $\frac{25 \times ?}{2.5} = \frac{40}{2.5}$

$? = 16$

PRACTICE

Circle the greater measurement in each pair.

1. 1 inch   1 centimeter
2. 1 gram   1 ounce
3. 1 quart   1 liter
4. 1 yard   1 meter
5. 1 mile   1 kilometer

Fill in each blank. Round all answers to the nearest tenth.

6. 20 yards ≈ __ meters
7. 40 grams ≈ __ ounces
8. 15 mi ≈ ______ km
9. 10 liters ≈ __ quarts
10. 3 meters ≈ __ yards
11. 15 lb ≈ ______ kg
12. 100 cm ≈ ______ in.
13. 150 gal ≈ ______ kL
14. 12 in. ≈ ______ cm

Adding and Subtracting Mixed Measurements

When two measurements are in the same units, you can add and subtract them like other numbers.

To add mixed units:
1. Add the smaller units.
2. Add the larger units.
3. Simplify your answer by converting as many of the smaller units as possible into the larger units.

1 hour	15 minutes
+ 2 hours	50 minutes
3 hours	65 minutes

= 3 hours + 60 minutes + 5 minutes
= 3 hours + 1 hour + 5 minutes
= 4 hours 5 minutes

To subtract mixed units:
1. Subtract the small and large units separately.
2. Simplify your answer.

In a subtraction problem, you may have to borrow.

~~3~~ 2 ft	~~3~~ 15 in.
− 1 ft	9 in.
1 ft	6 in.

3 is less than 9, so "borrow" 12 inches from 3 feet.

15 − 9 = 6
2 − 1 = 1
The difference is 1 ft 6 in.

PRACTICE

Add or subtract the measurements below. Then simplify your answer. You can use the tables on pages 111, 113, and 114.

1 2 hours 45 minutes
+1 hour 15 minutes

2 4 pounds 10 ounces
+ 5 pounds 7 ounces

3 5 yards 2 feet
+ 2 yards 2 feet

4 12 feet 10 inches
+ 3 feet 7 inches

5 1 hour 15 minutes
+ 50 minutes

6 1 quart 3 cups
+ 1 quart 2 cups

7 12 pounds 10 ounces
− 5 pounds 7 ounces

8 3 yards 2 feet
− 2 yards 1 foot

9 9 feet 4 inches
− 5 feet 8 inches

10 1 gallon 1 quart
− 3 quarts

11 4 hours 15 minutes
− 1 hour 45 minutes

12 4 pints
− 1 pint 1 cup

Finding Perimeter

The total length of the sides of a figure is called its **perimeter.**

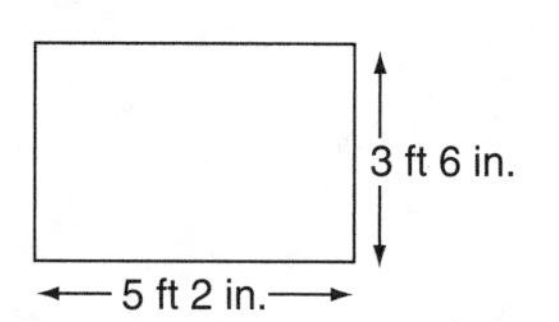	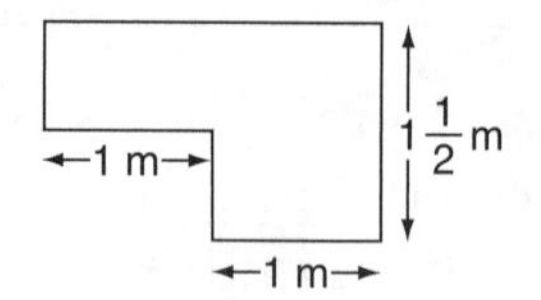	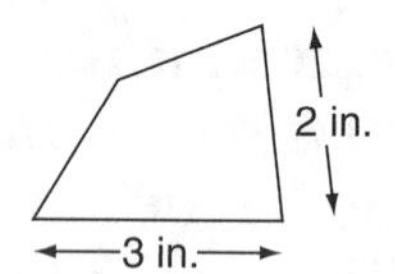
This figure is a rectangle, so opposite sides have the same length. The length of the top and bottom is 5 ft 2 in., and the length of the left and right edges is 3 ft 6 in. The perimeter is: 5 ft 2 in. 3 ft 6 in. 5 ft 2 in. + 3 ft 6 in. 16 ft 16 in. = 17 ft 4 in.	For this figure, the top edge is 1 meter + 1 meter = 2 meters. The right edge is $1\frac{1}{2}$ meters, so the sum of the two parts of the left edge is $1\frac{1}{2}$ meters. The perimeter is: $2 + 1\frac{1}{2} + 2 + 1\frac{1}{2} = 7$ meters	We do not know the lengths of the two unmarked sides. So we do not have enough information to calculate the perimeter of the figure.

PRACTICE

Find the perimeter of each shape. *Hint:* The symbol " is an abbreviation for inches and the symbol ' is an abbreviation for feet.

1 A room is 15 feet long by 12.5 feet wide.

Perimeter: _______

2

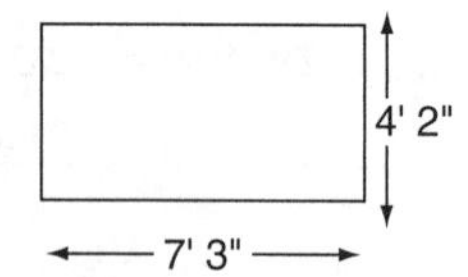

Perimeter: _______

3

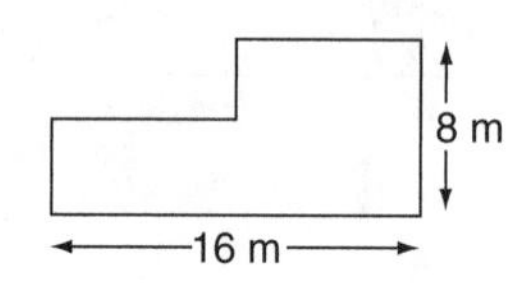

Perimeter: _______

4

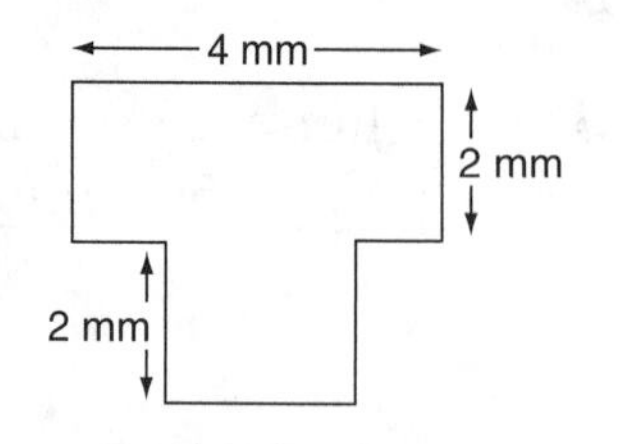

Perimeter: _______

5

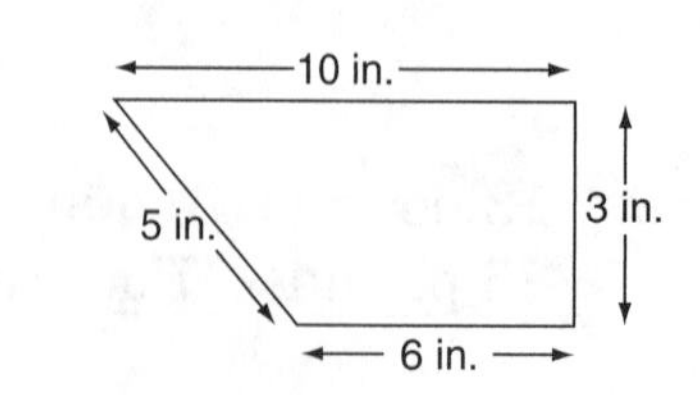

Perimeter: _______

6

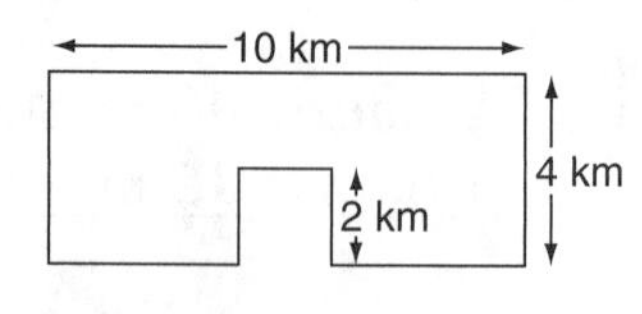

Perimeter: _______
(*Hint:* The "notch" along the bottom adds length to the perimeter.)

7 The perimeter of a square is 16 inches. How long is each side of the square?

____

Finding Area

The amount of a flat region inside a figure is called its **area.** To find the area of a square or a rectangle, you can multiply length times width.

Area = 6.5 yards × 2 yards
= 13 square yards
or
13 yd^2

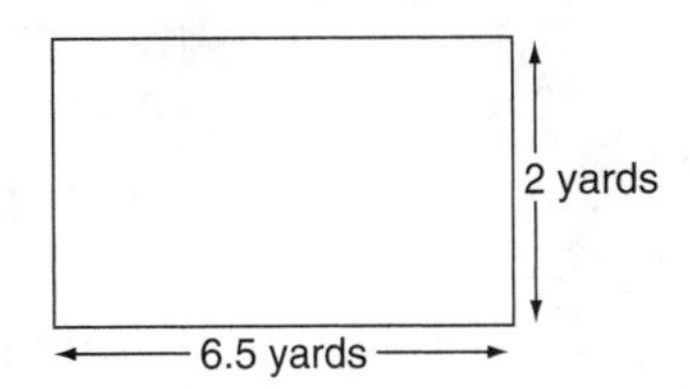

To find the surface area of a box, find the area of each of the six **faces** of the box. Then add those six measurements.

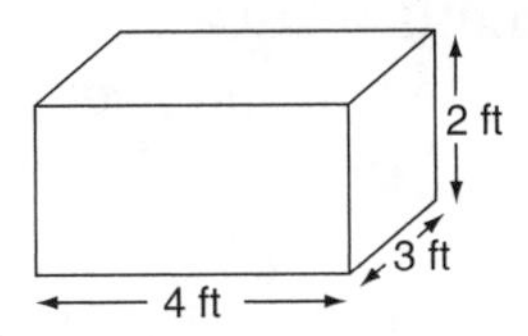

Area of the top	=	4 feet × 3 feet	=	12 ft^2
Area of the bottom	=	4 feet × 3 feet	=	12 ft^2
Area of the front	=	4 feet × 2 feet	=	8 ft^2
Area of the back	=	4 feet × 2 feet	=	8 ft^2
Area of the left face	=	3 feet × 2 feet	=	6 ft^2
Area of the bottom	=	3 feet × 2 feet	=	6 ft^2
Total Surface Area	=			52 ft^2

The abbreviation "yd^2" is pronounced "square yards," and the abbreviation "ft^2" is pronounced "square feet." Area is always measured in *square units.*

PRACTICE

Find the area or surface area of each shape.

1 a patio 5 meters by 5 meters

Area: _______

2 a wall 10 ft tall and 15 feet wide

Area: _______

3 a square end table with sides 0.5 yards long

Area: _______

4

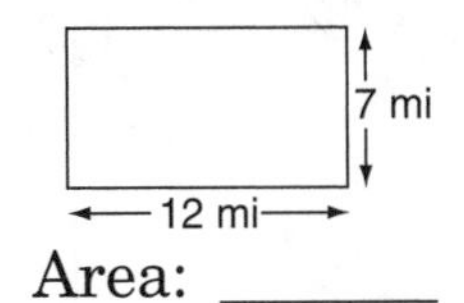

Area: _______

5

18"
5"

Area: _______

6

3 ft
2 ft

Area of the shaded section:

____

7

57 mm
a square

Area: _______

8

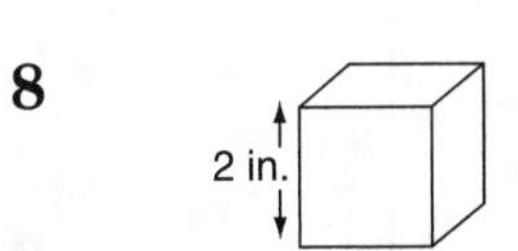

This is a cube. Its six faces are identical squares.

Surface Area: _______

9 A room is 20 m by 10 m, with walls that are $8\frac{1}{4}$ m high. To paper all four walls, how many square meters of wallpaper do you need? (Ignore doors and windows.)

Finding Volume

The amount of space inside a three-dimensional figure is called its **volume.** To find the volume of a rectangular box, you can multiply length times width times height.

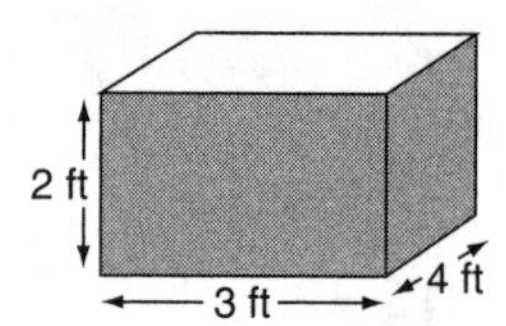

Volume = length × width × height.
Volume = 3 ft × 2 ft × 4 ft = 24 ft^3

Remember: to multiply three numbers, find the product of the first two. Then multiply that product by the third number.

Volume is measured in **cubic units,** or units^3. One cubic foot means a cube that is one foot long on each side. Similarly, one cubic centimeter means a cube that is 1 centimeter long on each side.

PRACTICE

Find the volume of each box.

1

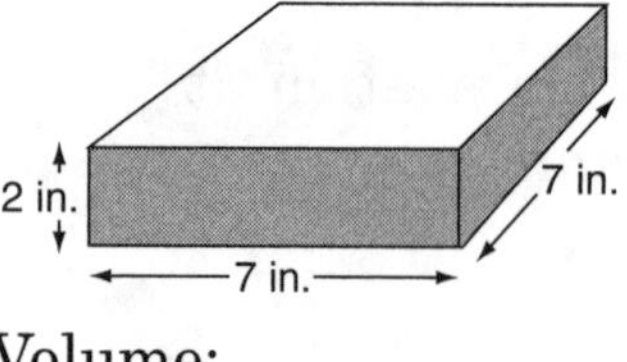

Volume: ______

2

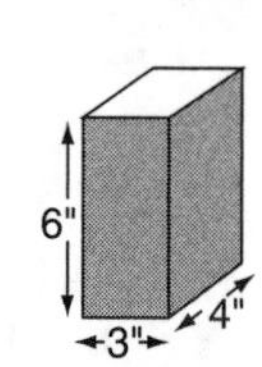

Volume: ______

3

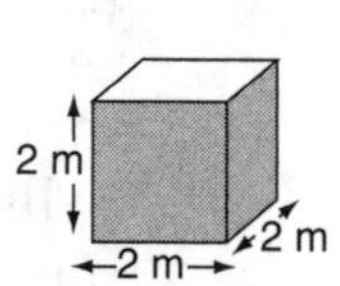

Volume: ______

4

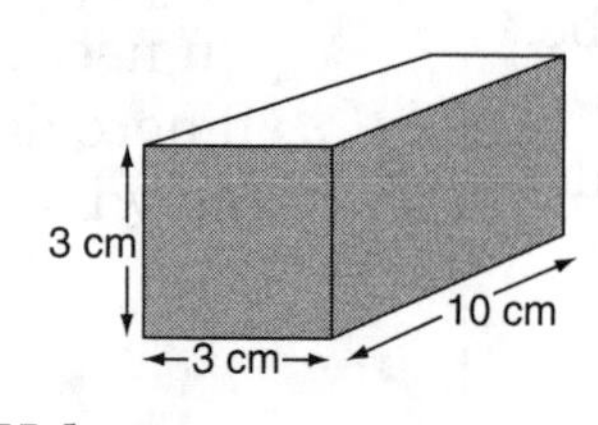

Volume: ______

5 Anthony needs 1 cubic foot of aquarium gravel. That would be enough to fill what fraction of this figure?

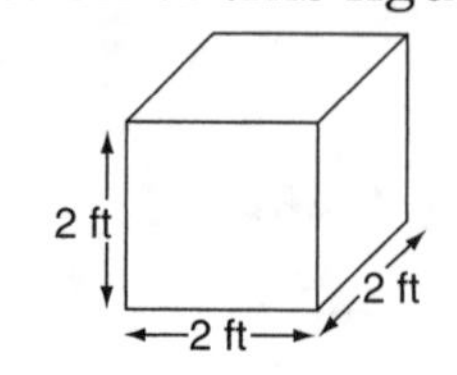

______

6

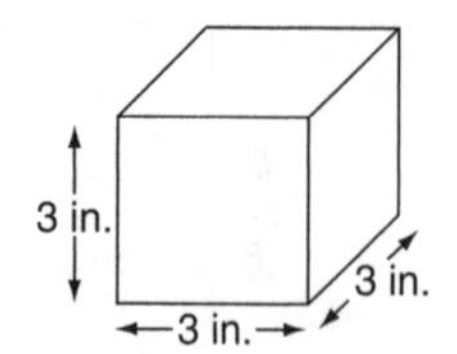

The length of the cube is 3 inches. What is the volume of this cube?

______

7 Erica bought 6 m^3 of gravel. Which dimension represents a bin that will hold all the gravel?

A 2 m × 2 m × 1 m
B 1 m × 1 m × 5 m
C 2 m × 2 m × 2 m

8 Erica's wheelbarrow has a carrying box whose size is 0.5 m × 1 m × 1 m. To move all 6 cubic meters of gravel, how many times will she have to fill the wheelbarrow?

______

Changes in Time

For the problem below, imagine how the hands on the clock move. Count the number of *complete* hours that pass. Then count the minutes that pass.

Problem
Eric started work at 8:25. He finished at 4:10. How long did he work?

Count the hours: The number of *complete* hours from 8:25 to 3:25 is 7.
Count the minutes:
From 3:25 to 4:10 is 45 minutes.

Answer: 7 hours 45 minutes

For the problem below, you want to find when you should start a task so you will finish at a particular time. To do this, count the change in hours first. Then add the change in minutes.

Problem
It takes 1 hour 15 minutes to drive to a doctor's office. You have a 10:00 appointment. When should you leave?

Hint: You need a time *before* 10:00, so count back from 10:00.

Count the hours:
1 hour before 10:00 is 9:00.
Count the minutes:
15 minutes before 9:00 is 8:45.
Answer: You should leave at 8:45.

Pay close attention to any mention of A.M. (morning) and P.M. (afternoon and evening). From 9:00 A.M. to 10:00 A.M. is 1 hour, but from 9:00 A.M. to 10:00 P.M. is 13 hours.

PRACTICE

Fill in the blank for each problem.

1 What time is 3 hours *before* 1:15? ______

2 It will take $3\frac{1}{2}$ hours to make dinner rolls. You plan to eat at 12:15 P.M. When should you start the rolls?

3 You need to work $5\frac{1}{2}$ hours today. If you start at 11:15, when will you finish? ______

4 Your night class starts at 6:55 and ends at 7:35. How long is the class? ______

5 It takes you 20 minutes to clean one room, and there are 3 rooms in your apartment. If you start cleaning at 8:10, when will you finish? ______

Measurement Skills Practice

Circle the letter for the correct answer to each problem.

This diagram is from the plans for building a bookcase. Study the diagram. Then do Numbers 1 through 6.

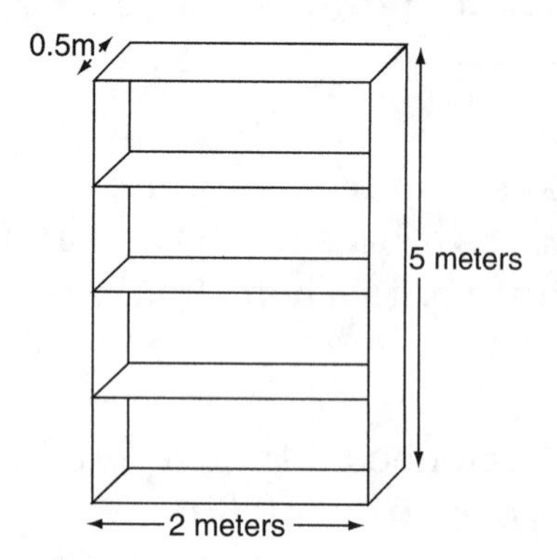

1 Each shelf on the bookcase will be 0.5 (or $\frac{1}{2}$) meter deep. Which of these is another way to describe the depth of each shelf?

A 5 centimeters
B 20 centimeters
C 50 centimeters
D 95 centimeters

2 The top of each shelf in this bookcase will be covered with a special adhesive-backed paper. How much of the paper is needed to cover four shelves?

F 1 square meter
G 4 square meters
H 8 square meters
J 10 square meters

3 This bookcase covers a region of the wall. What is the *perimeter* of that region?

A 7 meters
B 14 meters
C 10 meters
D 20 meters

4 According to the plans, each shelf on this bookcase can hold 110 kilograms. How many pounds can each shelf hold? (One kilogram is about 2.2 pounds.)

F 5
G 41
H 242
J 24.2

5 Akira estimates that it will take him $7\frac{1}{2}$ hours to build this bookcase. If he works for 90 minutes a day, how many days will it take him to finish?

A 4 days
B 5 days
C 6 days
D 9 days

6 Akira needs $\frac{1}{2}$ gallon of shellac to finish the bookcase, but the shellac only comes in 1-quart cans. How many 1-quart cans does he need?

F 1
G 5
H 4
J 2

This diagram shows plans for two flower beds Helio is going to put along the sides of his driveway. Study the diagram. Then do Numbers 7 through 13.

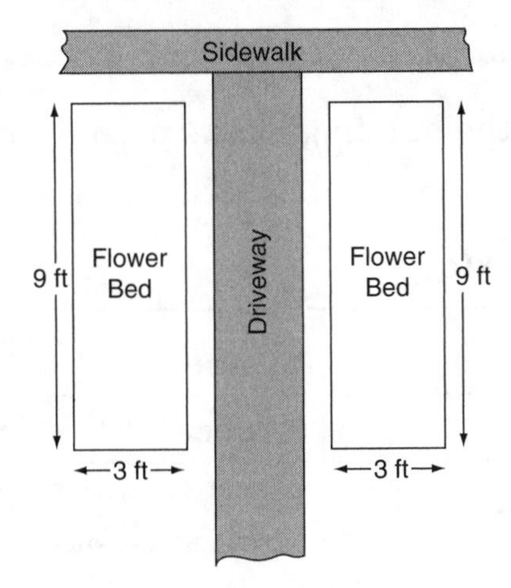

7 Helio wants to put cedar planks around the outside edge of each flower bed. To figure out how much planking he needs for each flower bed, what should he find?

A the perimeter
B the area
C the volume
D the depth

8 Each flower bed will be raised 2 feet above the ground. How much soil does Helio need to fill each flower bed?

F 27 feet^3
H 28 feet^3
G 44 feet^3
J 54 feet^3

9 Helio begins building the flower beds at 8:25 A.M. He finishes at 2:15 P.M. How much time did he spend on this task?

A 5 hours 50 minutes
B 6 hours 50 minutes
C 5 hours 90 minutes
D 5 hours 10 minutes

10 Helio should use an ounce of fertilizer per square feet of area. How much fertilizer does he need to cover both flower beds?

F 48 ounces
G 24 ounces
H 54 ounces
J 108 ounces

11 Helio buys $\frac{1}{4}$ of a pound of fertilizer. How many ounces is that?

A 25
B 5
C 4
D 8

12 It takes 20 minutes for Helio to plant half of one flower bed. At that rate, how long will it take him to plant all the *remaining* portions of the flower beds?

F 1 hour 15 minutes
G 1 hour 10 minutes
H 2 hours
J 1 hour

13 Helio's garden hose can supply 5 gallons of water a minute. At that rate, how much water can it supply in an hour?

A 30 gal
B 50 gal
C 300 gal
D 60 gal

Geometry

Basic Concepts

Geometry is the study of shapes. The ideas below are the building blocks used to create and describe shapes.

Basic Ideas in Geometry

Name	Meaning	Diagram	Symbol
point	a dot; a single location		*A* or *B;* a single letter
line	a straight, connected set of points; it extends in two directions		two letters with a double arrow above, such as $\overleftrightarrow{AB}$
line segment	the part of a line that is between two endpoints		two letters with a bar above, such as $\overline{AB}$
angle	an amount of a turn		the symbol $\angle$ followed by three letters, such as $\angle ABC$

PRACTICE

Draw a figure for problems 1–4 below.

1 Draw a point and label it *C*.

2 Draw a line segment and label it with points *D* and *E*.

3 Draw a line and label it with points *S* and *T*.

4 Draw an angle and label it with points *M, O,* and *P*.

5 $\angle FGH$ is the symbol for a(n) _?_ labeled with the points *F, G,* and *H*.

____

6 $\overleftrightarrow{FG}$ is the symbol for a(n) _?_ containing points *F* and *G*.

____

7 How is a line segment different from a line?

Use this diagram to answer questions 8 and 9.

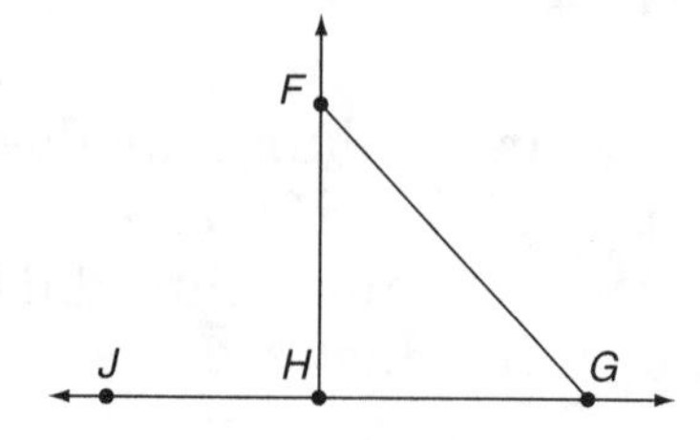

8 List all the labeled points in this diagram.

9 Name two of the angles shown in the diagram.

Angles

The **size** of an angle refers to the opening between the sides of the angle. The point where the sides meet is called the **vertex** of the angle.

The angle at the left is greater than the angle at the right.

You can think of an angle as part of a circle. A complete circle is 360 degrees (or 360°). At the right, look at the angle made by a square corner with its vertex at the center of the circle. That angle represents $\frac{1}{4}$ of a circle, so its measurement is $\frac{1}{4}$ of 360° or 90°.

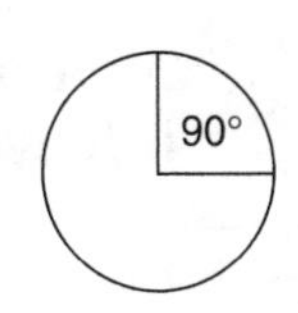

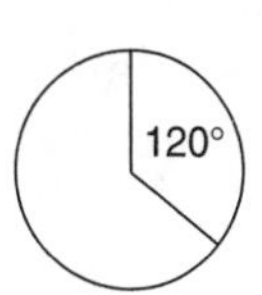

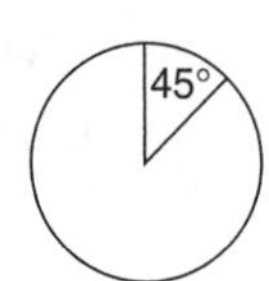

A full circle is 360°.

One-quarter of a circle is $\frac{360}{4} = 90°$.

One-third of a circle is $\frac{360}{3} = 120°$.

One-eighth of a circle is $\frac{360}{8} = 45°$.

PRACTICE

In each pair below, circle the letter for the smaller angle.

1 A 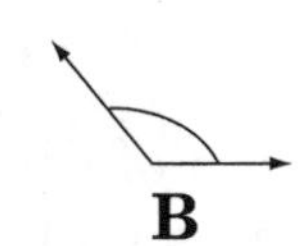 B

2 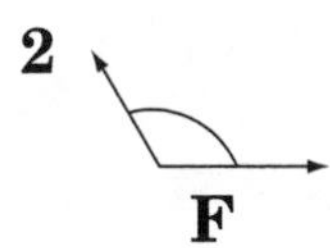 F 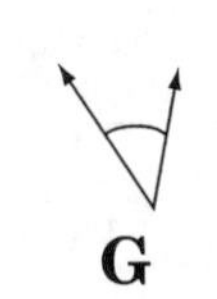 G

3 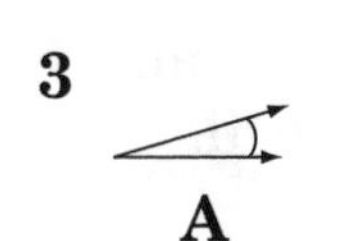 A 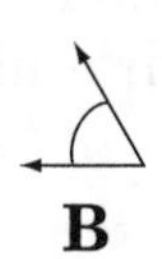 B

4 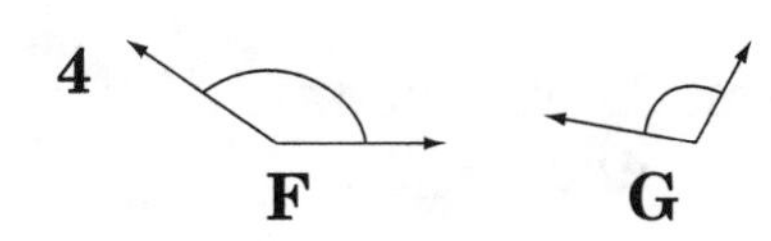 F G

For each angle, circle the letter for the best estimate of its measurement. If you have a protractor, use it to find the actual measurement of the angle.

5 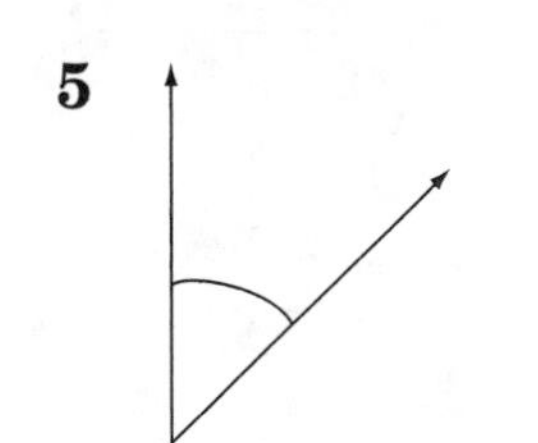

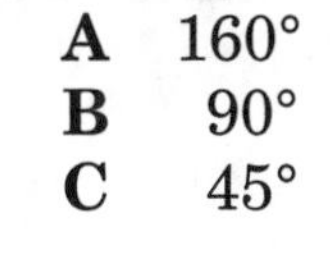

A 160°
B 90°
C 45°

6 

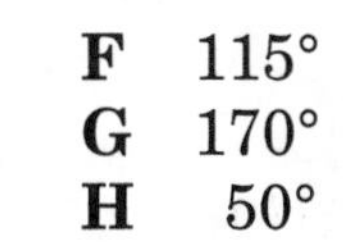

F 115°
G 170°
H 50°

7 

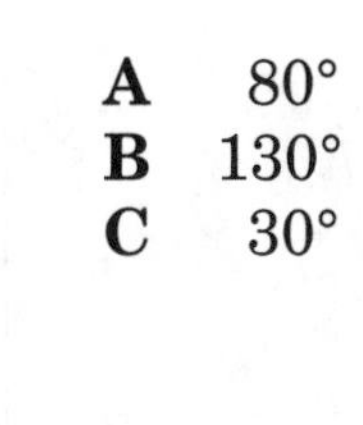

A 80°
B 130°
C 30°

Types of Angles

An angle whose measurement is 90° is called a **square corner** or a **right angle.** An angle smaller than a right angle is called an **acute angle.** An angle greater than a right angle (but less than 180°) is called an **obtuse angle.**

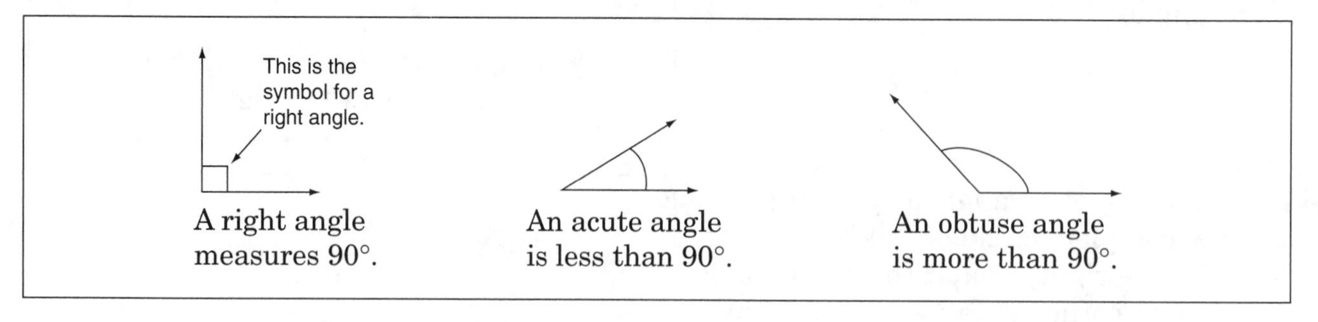

PRACTICE

For each angle below, write whether it is acute, right, or obtuse. You may use the square corner of a piece of paper to help you decide.

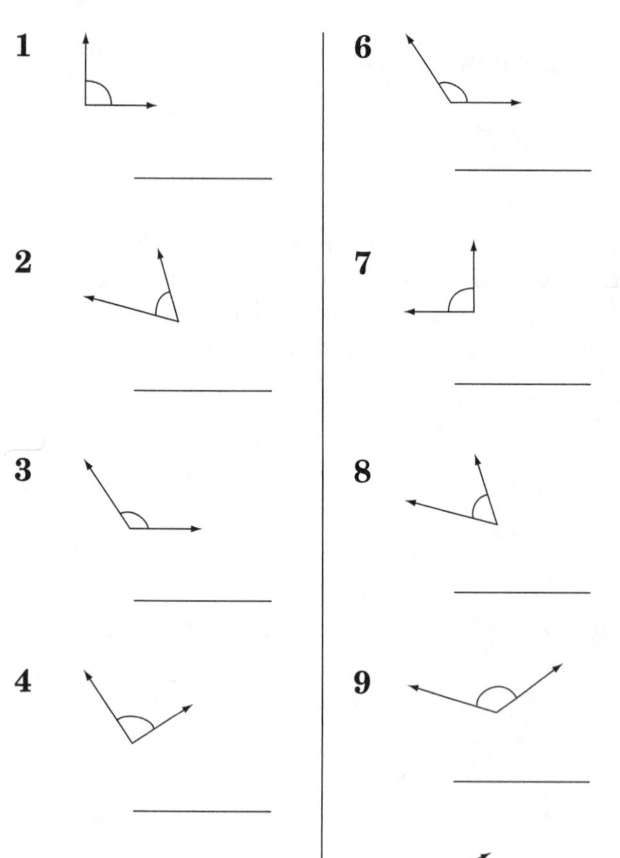

Use this diagram to answer questions 11 through 13.

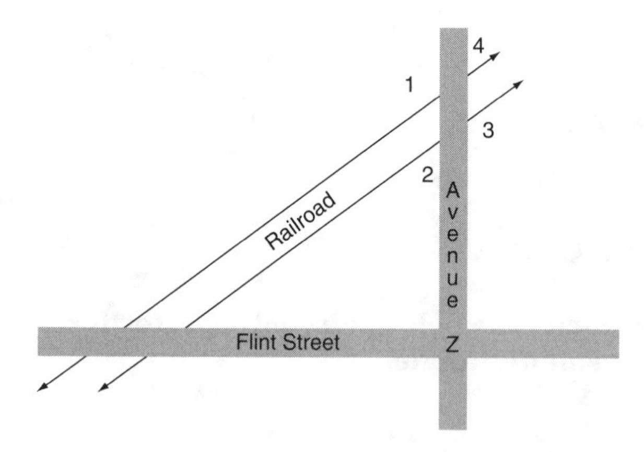

11 The intersection of Flint Street and Avenue Z creates four __?__ angles.

____

12 The angles formed by the intersection of the railroad and Avenue Z have been labeled with numbers. List all the acute angles that are formed.

____

13 On the diagram, the railroad ties would be represented as __?__.

A angles
B line segments
C points
D lines

Lines

Two lines that cross or that will cross are called **intersecting lines.**

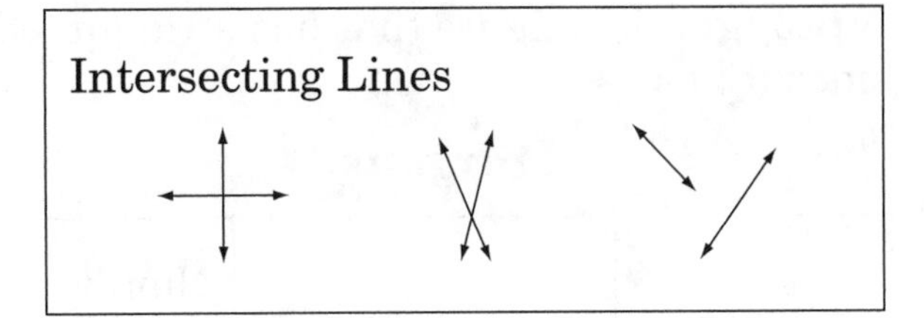
Intersecting Lines

Two lines that are always the same distance apart are called **parallel lines.**

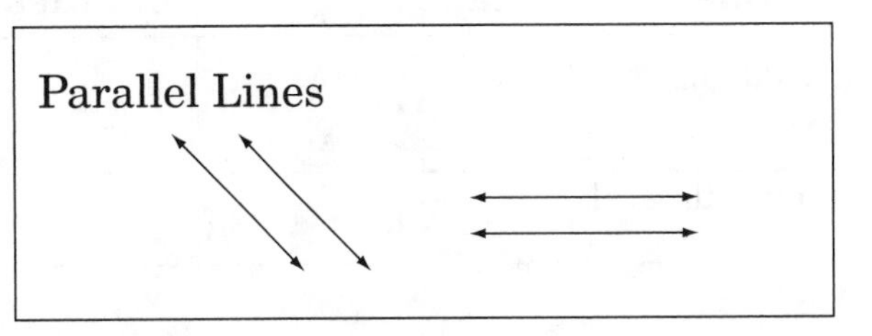
Parallel Lines

Two lines that form right angles when they meet are called **perpendicular lines.**

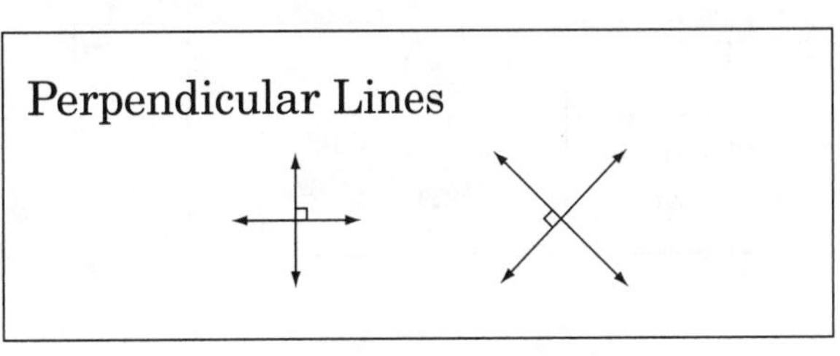
Perpendicular Lines

PRACTICE

Tell whether each pair of lines is parallel or intersecting.

1 ______

2 ______

3 ______

4 ______

Does each diagram show perpendicular lines? Write "yes" if the lines are perpendicular. Write "no" if they are not perpendicular.

5 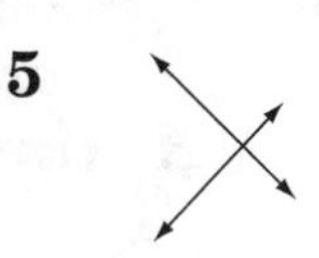 ______

6 ______

7 ______

8 ______

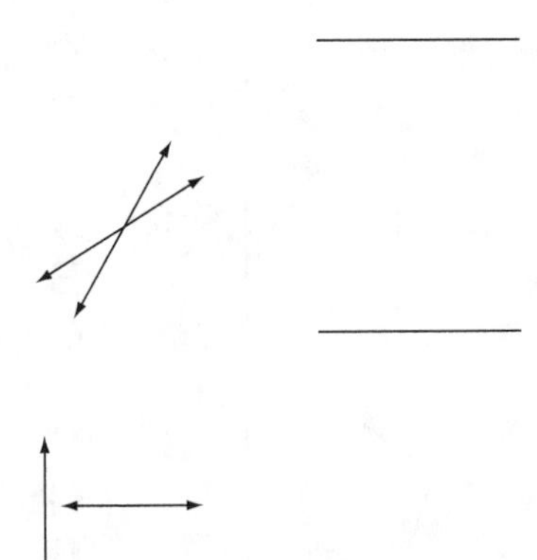

Use this diagram to answer questions 9 through 11.

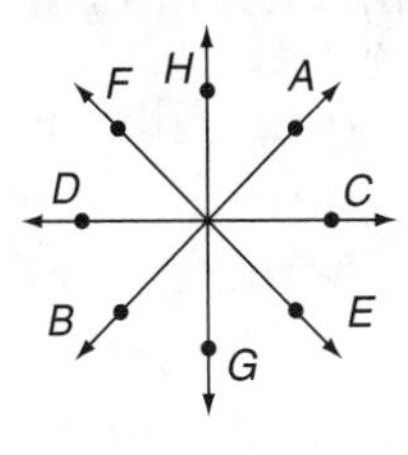

9 Which line or lines are perpendicular to $\overleftrightarrow{AB}$?

______

10 Is $\overleftrightarrow{FE}$ perpendicular to $\overleftrightarrow{DC}$?

______

11 Are there any parallel lines in this figure? If so, name the lines.

______

Polygons

A **polygon** is a figure that has straight sides. A polygon is named for the number of sides it has.

Polygons

Name	Shape	Number of Sides
triangle	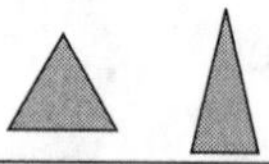	3
quadrilateral	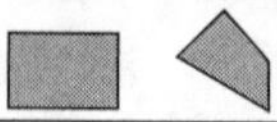	4
pentagon	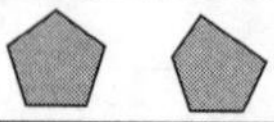	5
hexagon	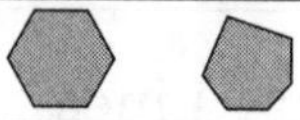	6
octagon	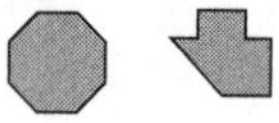	8

Special Types of Polygons

Name	Examples	Meaning
regular polygon	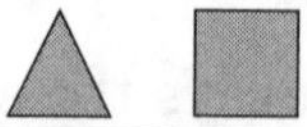	All the sides are the same length.
parallelogram		A parallelogram has four sides. The opposite sides are parallel.
rectangle	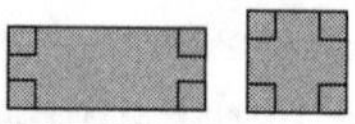	A rectangle has four right angles.
square		A square has four right angles and four equal sides.
trapezoid		A trapezoid has four sides. It has exactly one pair of parallel sides.

PRACTICE

There are three terms next to each figure. One of the three terms *does not* describe the figure. Circle the letter for the term that does not apply.

1 **A** quadrilateral
B parallelogram
C square

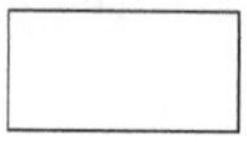

2 **F** parallelogram
G regular
H hexagon

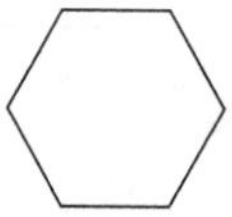

3 **A** quadrilateral
B parallelogram
C rectangle

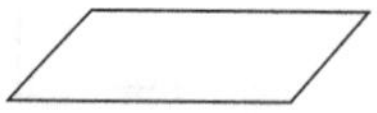

4 **F** trapezoid
G parallelogram
H rectangle

5 **A** pentagon
B regular
C octagon

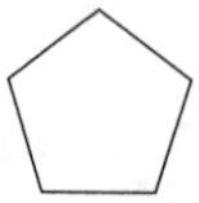

6 **F** quadrilateral
G parallelogram
H trapezoid

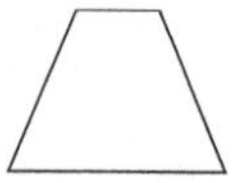

7 **A** polygon
B regular
C triangle

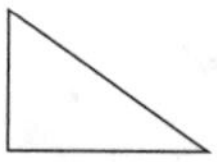

8 **F** triangle
G regular
H parallelogram

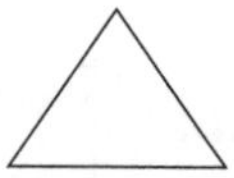

9 **A** polygon
B regular
C quadrilateral

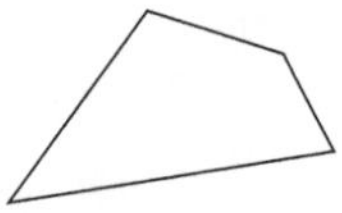

10 **F** pentagon
G hexagon
H regular

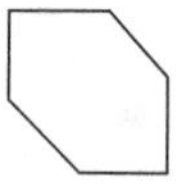

Types of Triangles

Triangles are a particularly interesting and important type of figure. For the sides, the sum of any two sides must be greater than the third side.

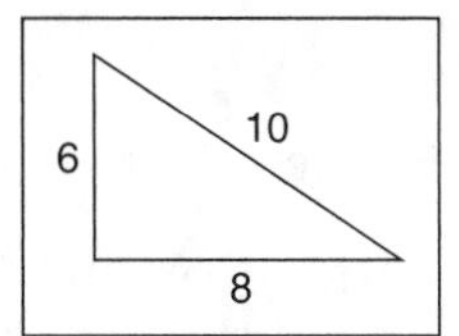

6 + 8 is greater than 10.
6 + 10 is greater than 8.
8 + 10 is greater than 6.

Here is another important thing to know about all triangles:

> If you add the measurements of all three angles in a triangle, the sum is always 180°.

Some triangles have three equal sides and three equal angles. Some triangles have two equal sides and two equal angles. For some triangles, the sides are three different lengths and the angles are three different sizes.

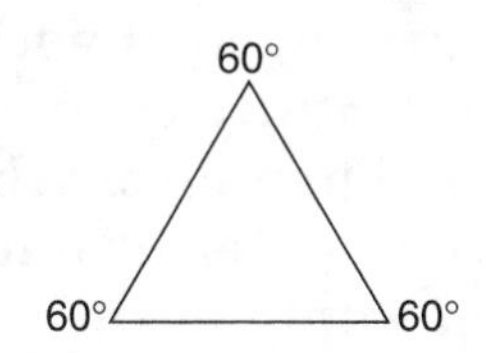

Equilateral triangle
All sides are equal.
All angles are equal.
$60° + 60° + 60° = 180°$

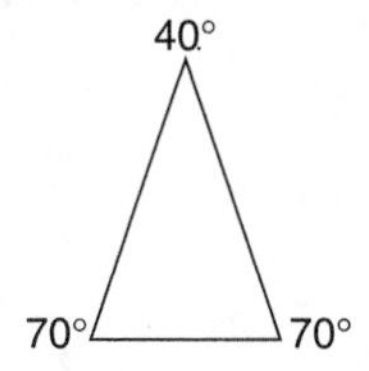

Isosceles triangle
Two sides are equal.
Two angles are equal.
$70° + 70° + 40° = 180°$

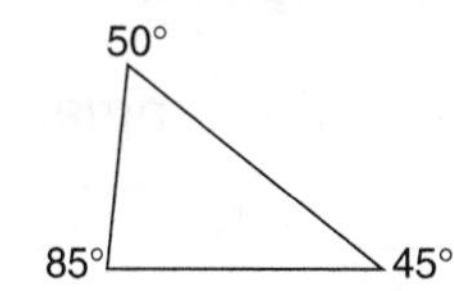

Scalene triangle
No two sides are equal.
No two angles are equal.
$85° + 45° + 50° = 180°$

PRACTICE

Use the classifications above to name each type of triangle. Then find out the measure of the unlabeled angle.

1 Type of triangle:

____

Measure of the unlabeled angle:

______

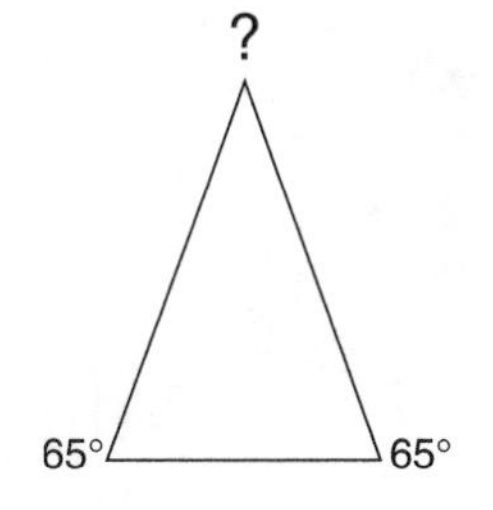

2 Type of triangle:

____

Measure of the unlabeled angle:

______

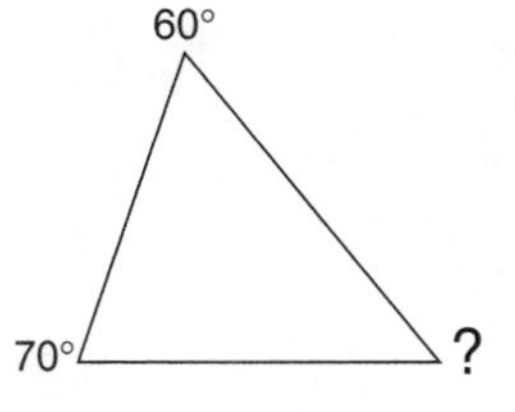

3 Type of triangle:

____

Measure of the unlabeled angle:

______

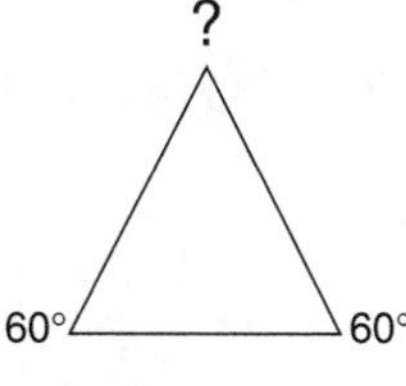

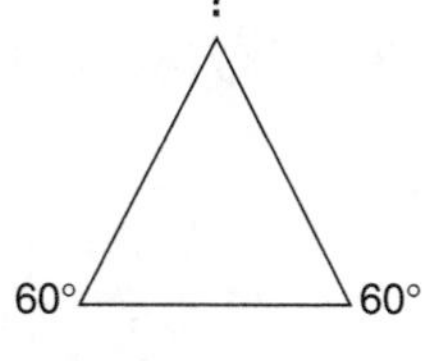

4 Type of triangle:

____

Measure of the unlabeled angle:

______

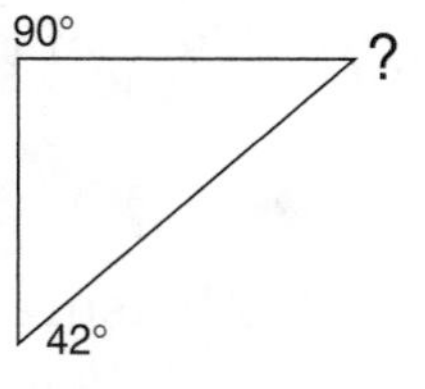

A different way to classify a triangle is to look at the size of the largest angle in the triangle.

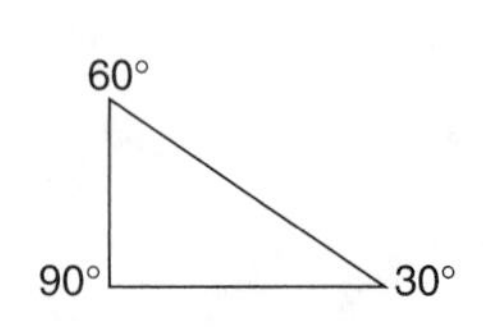

Right triangle
The largest angle in the triangle is a right angle.

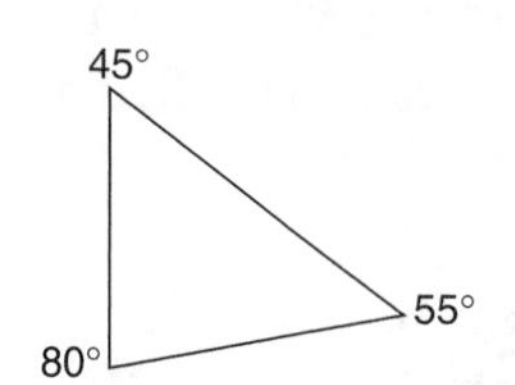

Acute triangle
The largest angle in the triangle is smaller than a right angle.

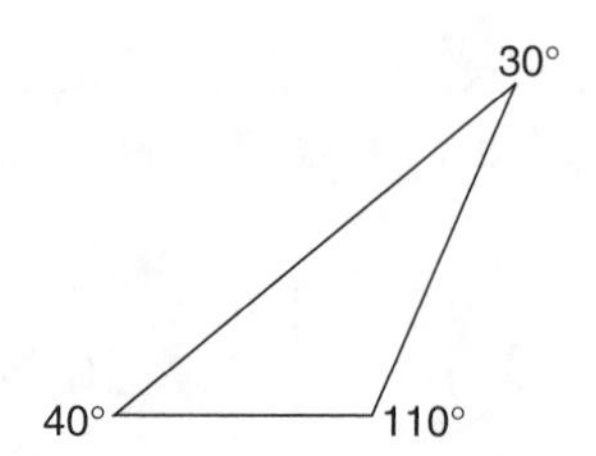

Obtuse triangle
The largest angle in the triangle is larger than a right angle.

PRACTICE

For each triangle below, tell whether the triangle is right, acute, or obtuse. Then find the measurement of the unlabeled angle.

5 Type of triangle:

____

Measure of the unlabeled angle: _______

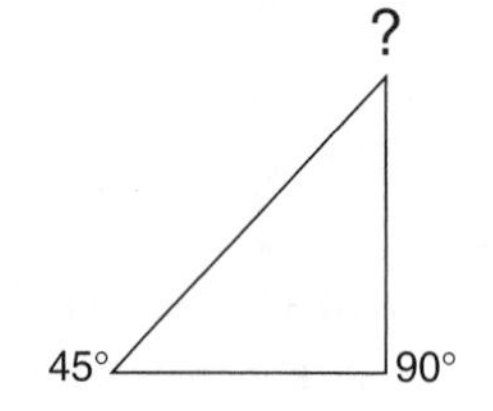

6 Type of triangle:

____

Measure of the unlabeled angle: _______

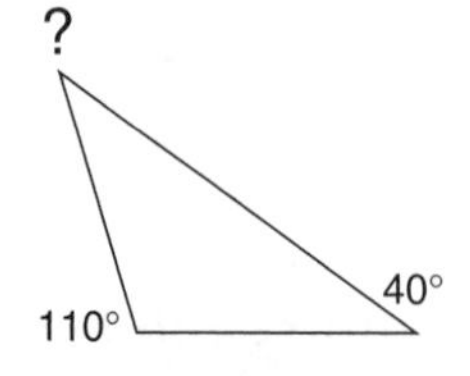

7 Type of triangle:

____

Measure of the unlabeled angle: _______

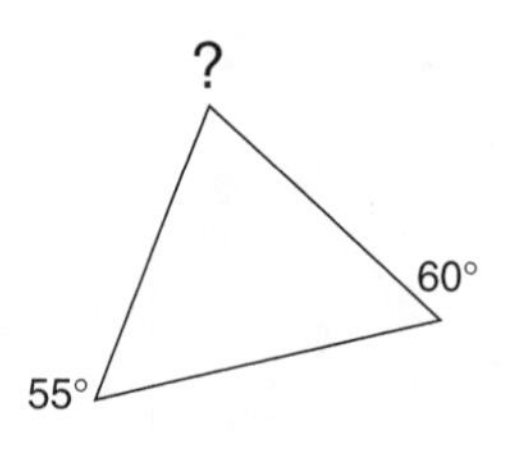

8 Type of triangle:

____

Measure of the unlabeled angle: _______

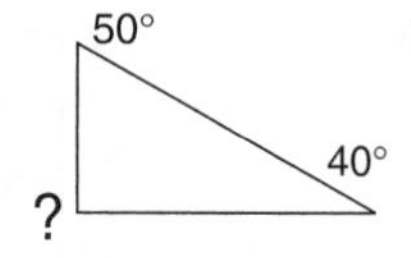

Below, two sides of a triangle that have the same length are marked with the symbol ||. Study each triangle. Then circle the letter for the term that *does not* apply to the triangle.

9 **A** right
B acute
C equilateral

10 **F** right
G scalene
H obtuse

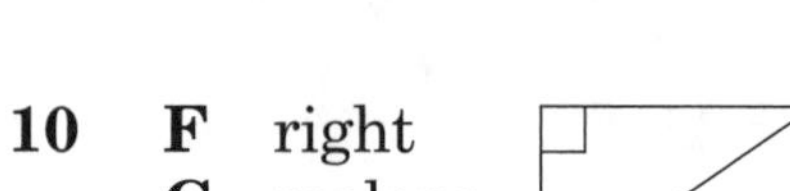

11 **A** obtuse
B isosceles
C scalene

12

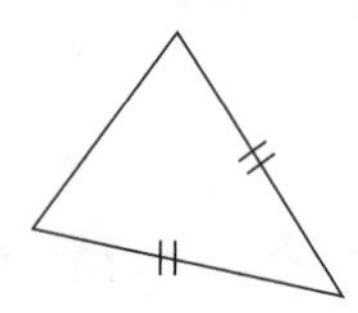

F acute
G isosceles
H equilateral

Circles

Another important shape is the circle. A **circle** is all the points that are a particular distance from a point called the **center** of the circle.

A circle contains all the points a given distance from its center point.

Imagine taping one end of a string to a piece of paper and tying a pencil to the other end. Stretch the string tightly and move the pencil. All the points will be on a circle.

The distance from the center of a circle to any point on it is called the **radius.** The width of the circle is called the **diameter,** and the diameter is twice as long as the radius.

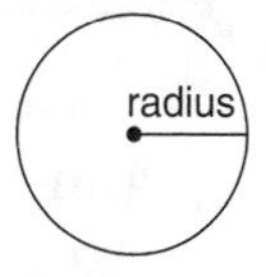

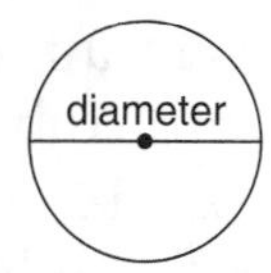

Notice that the diameter of a circle passes through the center of the circle.

PRACTICE

1

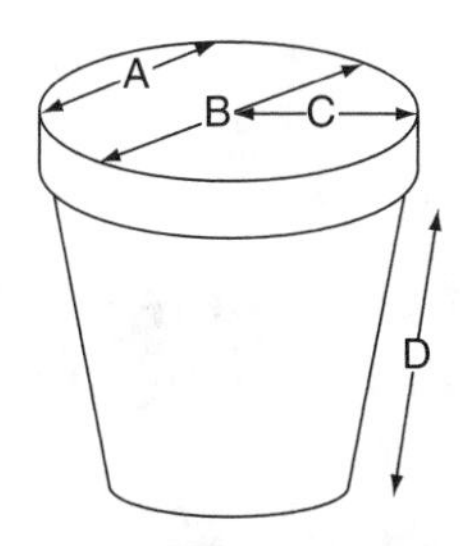

Jessica needs a flower pot that is 6 inches in diameter. Which line shows what she should measure?

____

2

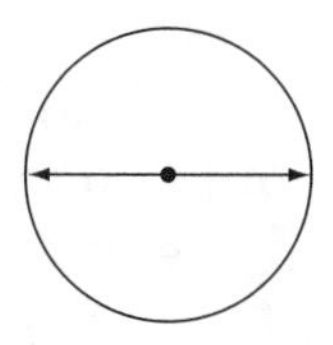

What do you call the widest distance across the top of the flower pot?

____

Find the diameter and the radius of each circle below. If you do not have enough information to do that, write "cannot tell."

3 Radius: _______

Diameter: _______

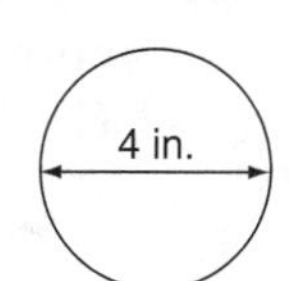

4 Radius: _______

Diameter: _______

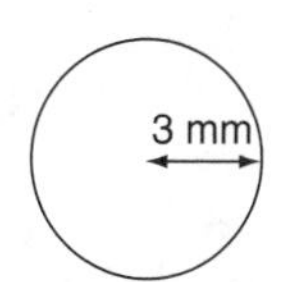

5 Radius: _______

Diameter: _______

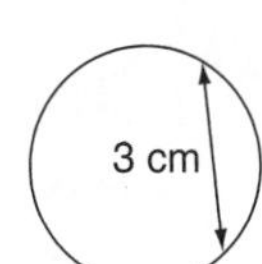

Finding the Circumference of a Circle

The distance around the outside of a circle is called its **circumference.** You can calculate the circumference of a circle if you know its diameter.

$$C = \pi d$$

The circumference of a circle is π times the diameter.

The value of π is about $\frac{22}{7}$. To two decimal places, the value of π is 3.14.

To find the circumference of any circle, multiply its diameter by 3.14. *Remember:* To find the number of decimal places in a product, add the number of decimal places in each of the two numbers being multiplied.

Problem: Hiram's hot tub is 4 meters wide. What is the circumference of the hot tub?

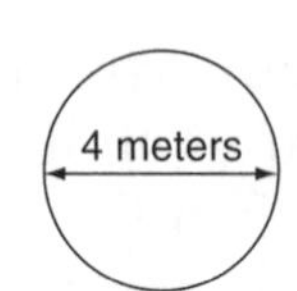

Solution:
Multiply 3.14 times 4.
There are two decimal places in 3.14 and 0 decimal places in 4, so the product has 2 decimal places.

$$\begin{array}{r} 3.14 \\ \times \quad 4 \\ \hline 12.56 \end{array}$$

PRACTICE

Below, find the circumference of each circle. Use the value 3.14 for π. *Hint:* Sometimes you are given the length of the diameter. Sometimes you are given the radius, and you must find the diameter.

1

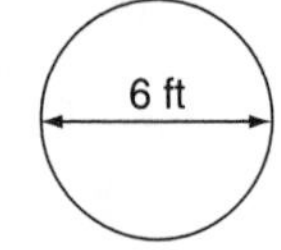

Circumference: ______

2

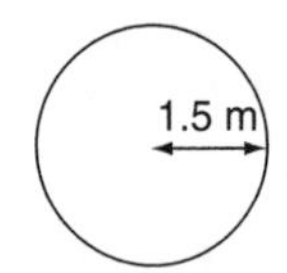

Circumference: ______

3

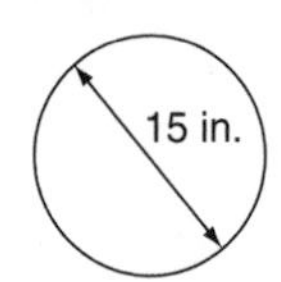

Circumference: ______

4 What is the circumference of a flower pot that is 8 inches in diameter? (Round your answer to the nearest inch.)

______

5 Brianna has a round table that is 2.5 yards in diameter. How much ribbon would she need to tack the ribbon all the way around the outside of the table?

______

6 Hope is making a wreath that is 40 centimeters in diameter. She needs enough ribbon to go around the wreath twice. How much ribbon does she need?

______

Recognizing Congruent Figures and Similar Figures

If two figures have the same size and the same shape, they are called **congruent** figures.

If two figures have the same shape, they are called **similar** figures. (Two similar figures *can be* the same size.)

PRACTICE

1 In the gray box, draw a figure that is congruent to the dark figure in the white box. Use the grid in the box to make sure your drawing is the same shape and the same size.

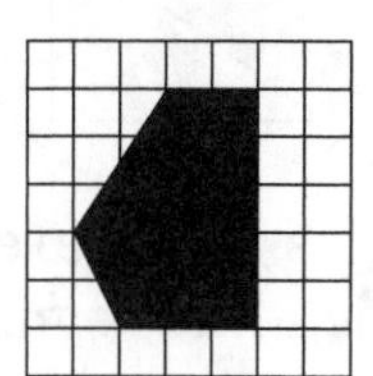
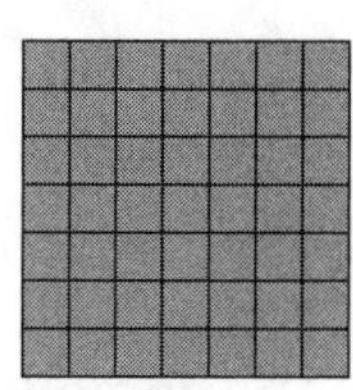

2 In the gray box, draw a figure that is similar to the dark figure but whose sides have half the lengths of the dark figure's sides.

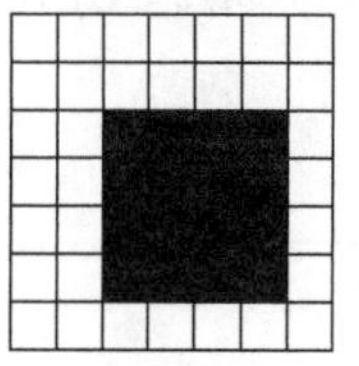
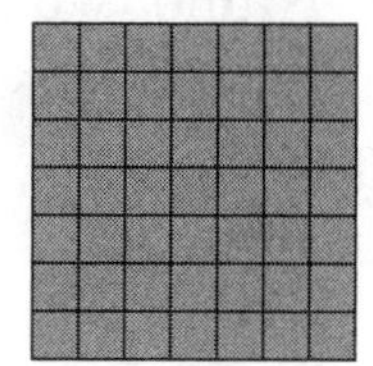

3 Circle the letter for the figure that is *congruent to* the figure in the white box.

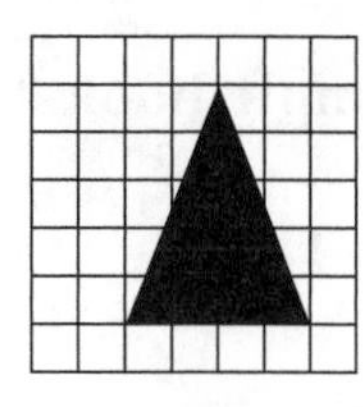
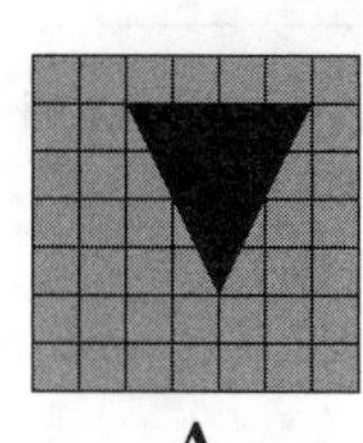
A
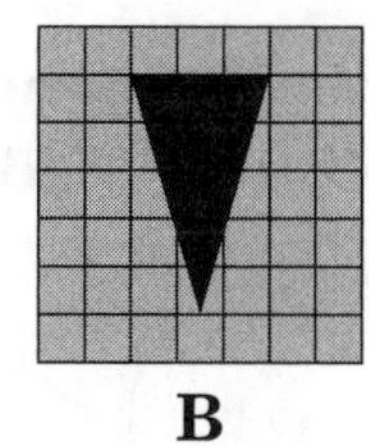
B
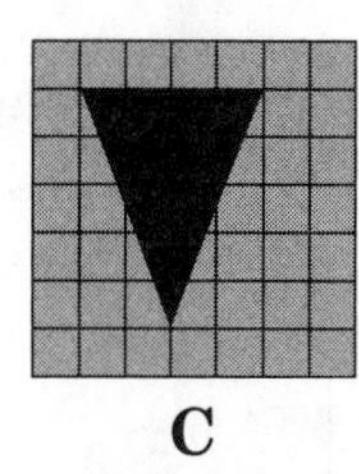
C

4 Circle the letter for the figure that is *similar to* the figure in the white box.

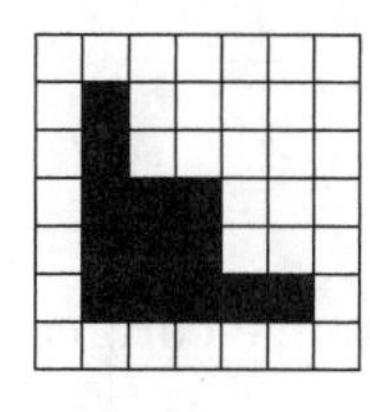
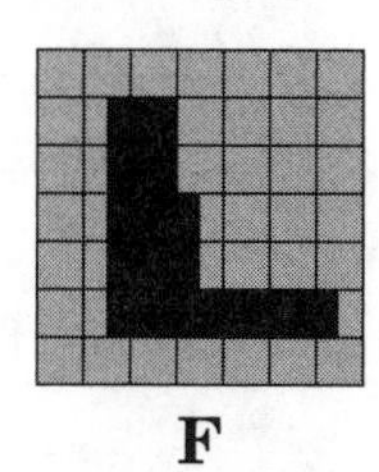
F
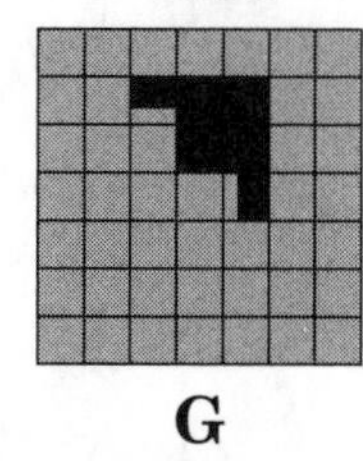
G
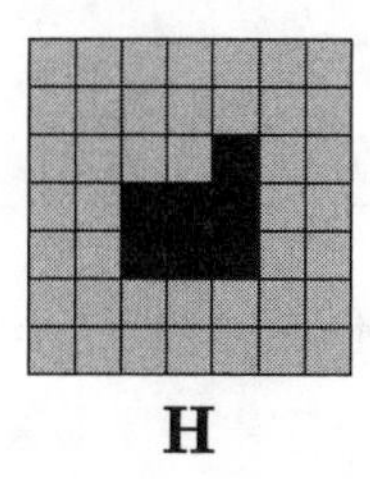
H

5 Circle the letter for the figure that is *congruent to* the figure in the white box.

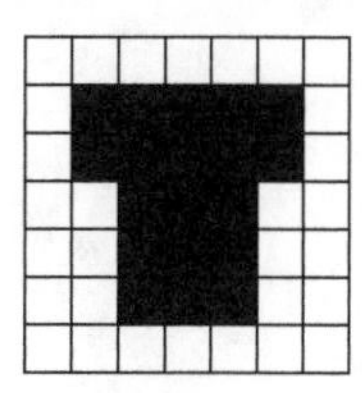
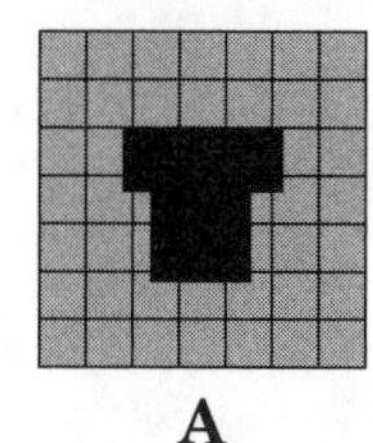
A
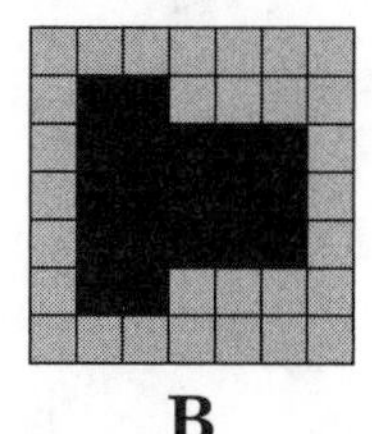
B
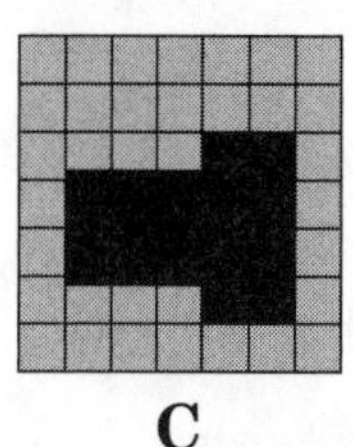
C

6

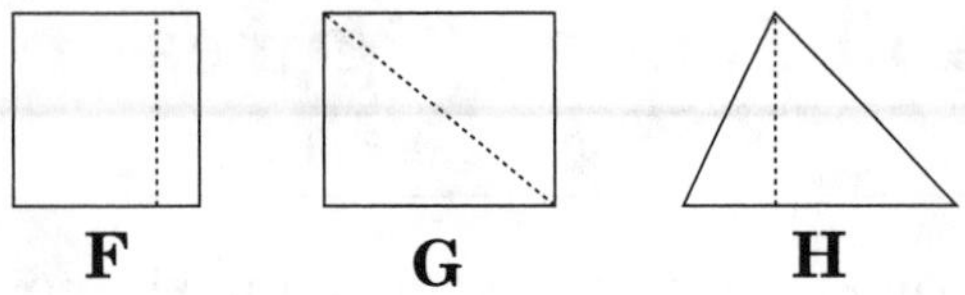

F G H

Circle the letter for the figure above that has been cut into two congruent halves.

7

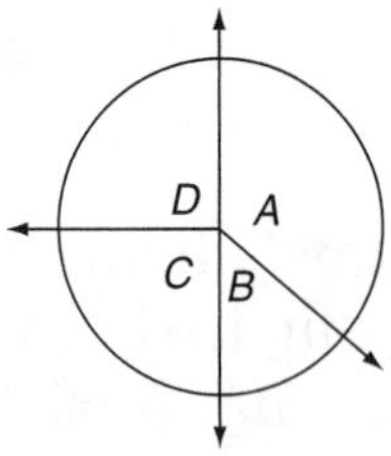

Name two congruent angles in this figure.

____

8

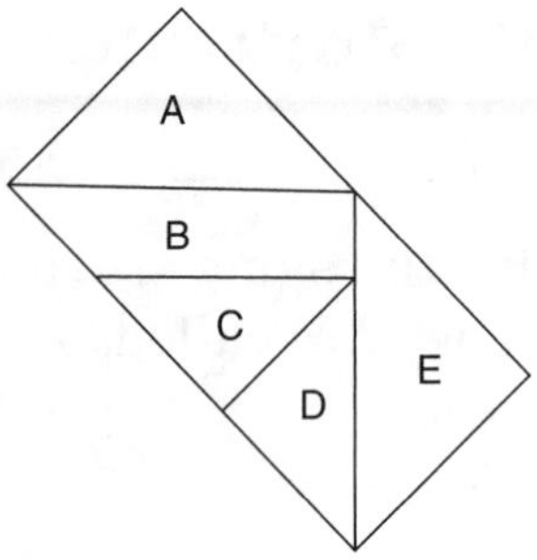

Which two sections of this figure are congruent?

____

9

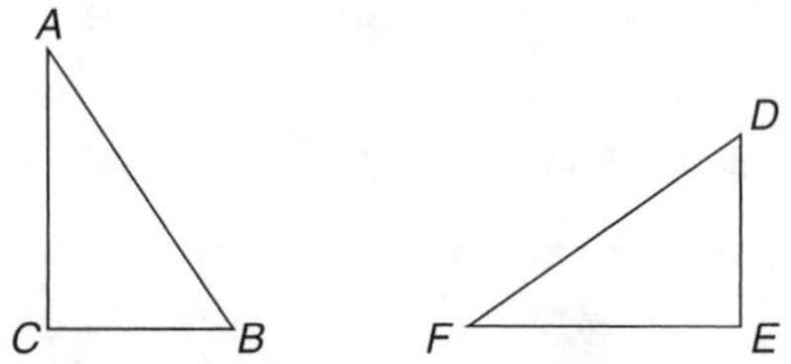

The two triangles above are congruent. Which line segment is the same length as segment *BC*?

____

Below, the diagram at the left shows the pieces needed to make the simple bookstand shown on the right. Use the diagram to answer questions 10 and 11.

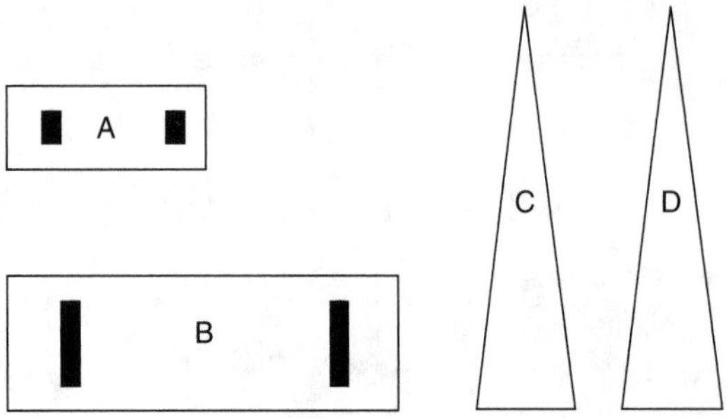

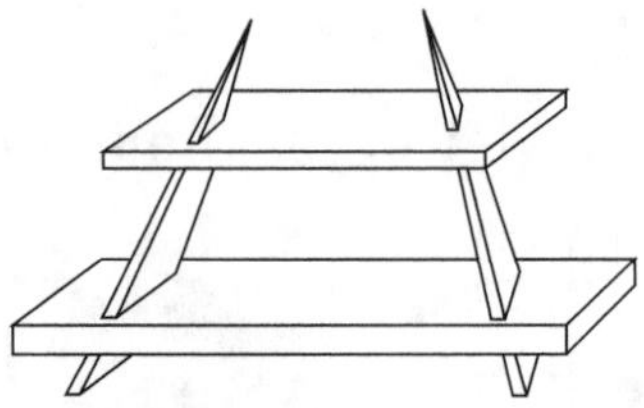

10 Which pieces, if any, are similar but not congruent?

____

11 Which pieces, if any, are congruent?

____

Three-Dimensional Figures

In the real world, objects have length, width, and height. The shapes of these objects are **three-dimensional figures,** or **solid figures.** The outsides of a three-dimensional figure are called its **faces.**

Common Solid Figures

A **cube** has six square faces. Every corner is a right angle.

A **rectangular solid** has six rectangular faces. Every corner is a right angle.

A **cylinder** is shaped like a tin can. The top and bottom faces are circles.

A **sphere** has the shape of a ball. A sphere consists of all points in space that are a particular distance from the **center** of the sphere.

A **cone** has one circular face. Opposite that circular face, the cone comes to a point called its **vertex.**

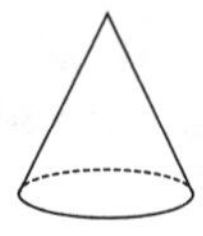

PRACTICE

Tell whether each object is shaped like a cube, a rectangular solid, a sphere, a cylinder, or a cone.

1 a water glass ______

2 a funnel minus the spout ______

3 a witch's hat ______

4 a section of pipe ______

5 bubbles in a carbonated drink ______

Use the definitions above to answer these questions.

6 A cube fits the definitions of both a cube and a(n) _?_. ______

7 *Yes* or *No*: Is this figure a rectangular solid?

______

8 Write the letter for the object that would hold more water.

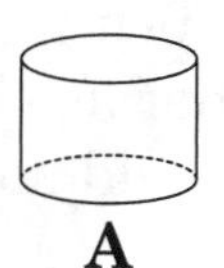

A

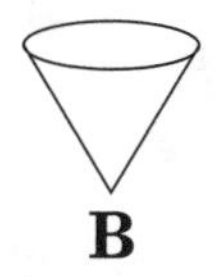

B

______

Visualizing Solids

An important skill in geometry is to draw or imagine figures. Also, you need to be able to look at figures from different points of view.

PRACTICE

For each problem on this page, draw or imagine a figure. Then use that image to solve the problem.

1 This cube is made up of smaller cubes. How many small cubes are there altogether in the large cube?

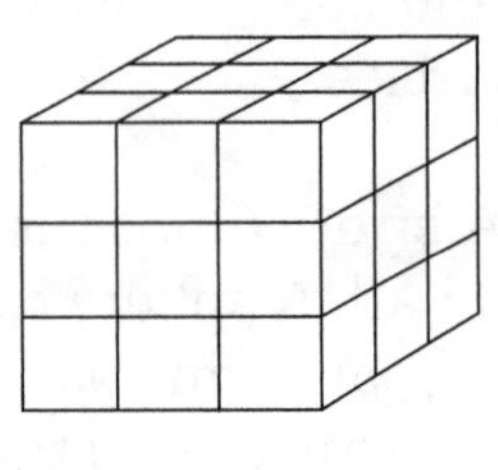

______

2 If you cut this cylinder in half along the dotted line, what is the shape of the cut ends?

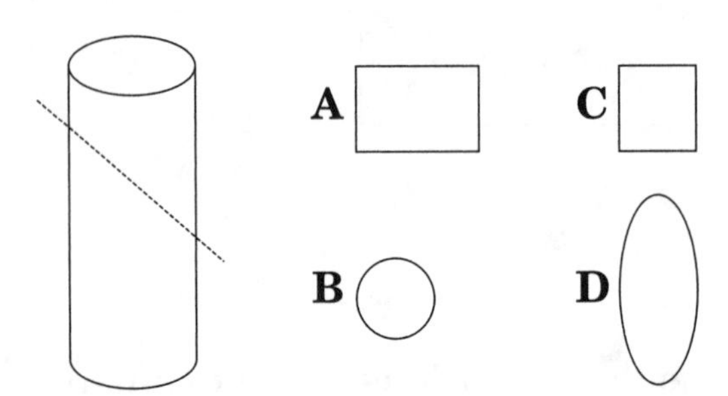

3

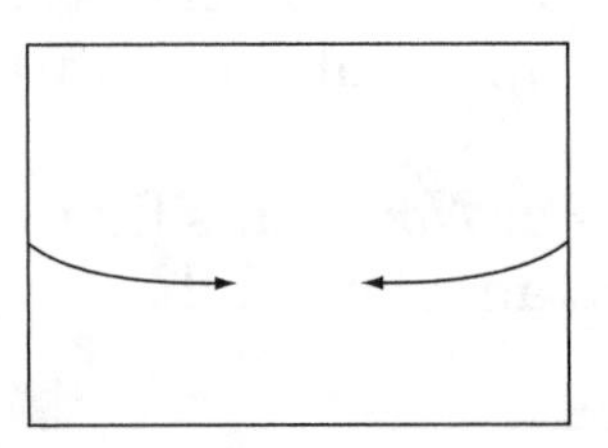

If you bring the sides of the paper together as shown by the arrows, what solid do you form?

F a cone
G a rectangular solid
H a cylinder
J a sphere

4 If you cut off the top of this cone along the dotted line, what is the shape of the cut end?

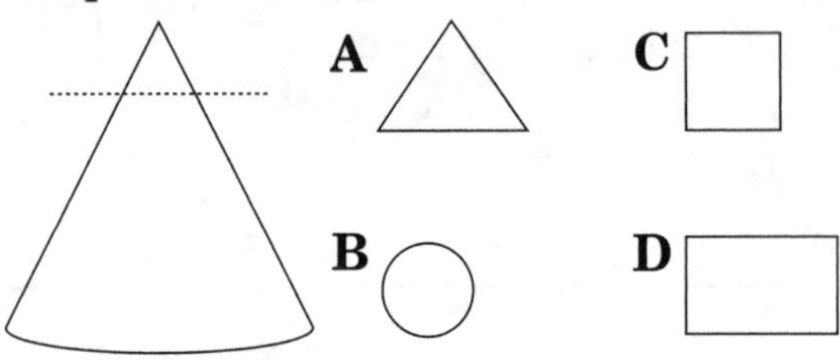

5 If you cut this cube in half along the dotted line, what is the shape of each of the two new pieces?

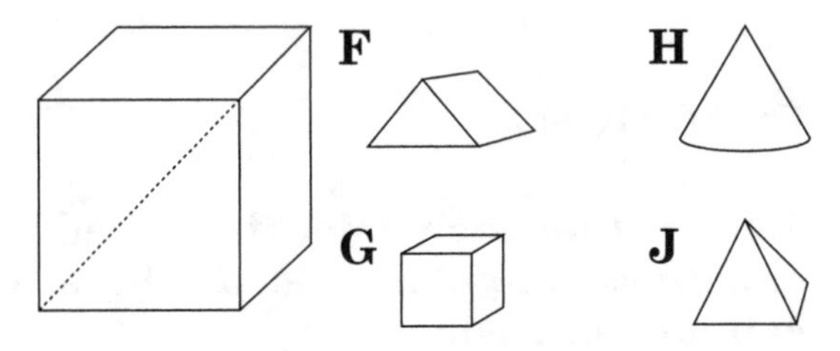

6 If you bring the sides of this figure together as shown by the arrows, what solid do you form?

A a cone
B a cylinder
C a rectangular solid
D a sphere

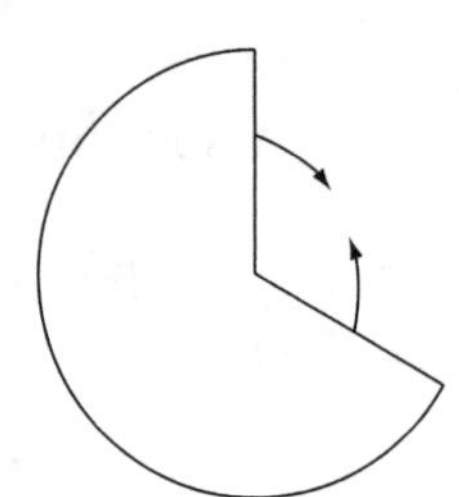

Geometry Skills Practice

Circle the letter for the correct answer to each problem.

This diagram shows the front of a fancy envelope. Study the diagram. Then do Numbers 1 through 7.

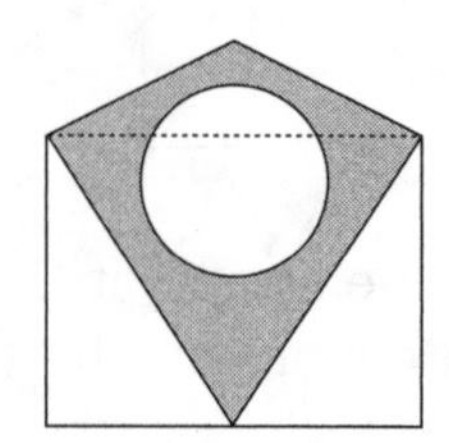

1 What shape is formed by the outside edges of this open envelope?

A a pentagon
B an octagon
C a hexagon
D a quadrilateral

2 Which parts of this diagram, if any, are congruent?

F the dark shape and the outside edges
G the circle and the dark shape
H the two triangles
J No two of the shapes are congruent.

3 What types of triangles are shown in the bottom corners of the diagram?

A equilateral
B isosceles
C right
D obtuse

4 The round seal near the middle of the envelope is two inches wide. What is its circumference? (Use 3.14 for π.)

F 3.14 inches
G 6.28 inches
H 1 inch
J This cannot be determined.

5 What type of angle is shown at the very top of the diagram?

A an acute angle
B a right angle
C a square corner
D an obtuse angle

6 Which line segments in the diagram, if any, are parallel?

F the bottom edge and the right edge
G the two top edges
H the right and left edges
J There are no parallel line segments in the diagram.

7 Which type of polygon is formed by the outside edges of the dark area?

A a regular polygon
B a quadrilateral
C a parallelogram
D all of the above

Below, the diagram at the left shows the fabric pieces needed to make the tent shown at the right. The sides of the tent are congruent triangles. Study the diagram. Then do Numbers 8 through 12.

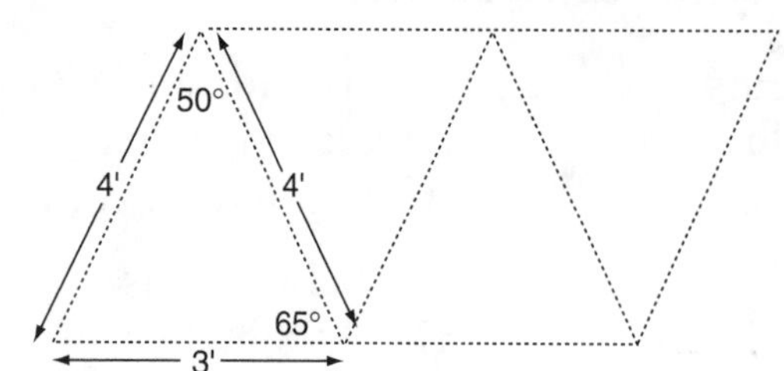

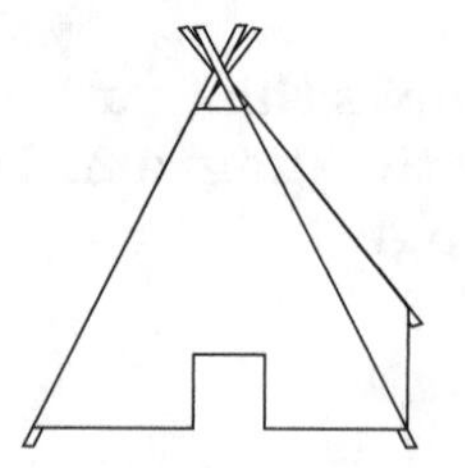

8 What type of line segments are formed by the tent poles?

F parallel
G perpendicular
H intersecting
J none of the above

9 What kind of triangle is formed by each side of the tent?

A equilateral
B isosceles
C right
D obtuse

10 What is the measure of the unmarked angle in the first triangle?

F 50°
G 100°
H 60°
J 65°

11 What is the shape of the region that this tent covers on the floor?

A a square
B a triangle
C a rectangle
D a circle

12 Each triangle in the diagram contains what type(s) of angles?

F obtuse only
G obtuse and acute
H acute and right
J acute only

Skills Inventory Post-Test

Part A: Computation

Circle the letter for the correct answer to each problem.

1

$245 \div 7 =$ _____

A 35
B 45
C 305
D 350
E None of these

2

$12.12 + 6.7 =$ _____

F 1.882
G 12.79
H 18.19
J 18.82
K None of these

3

$324 \times 21 =$ _____

A 9,720
B 6,804
C 972
D 6,704
E None of these

4

$4\overline{)856}$

F 216
G 214
H 224
J 211 r 2
K None of these

5

$$\begin{array}{r} 230 \\ \times\ 50 \\ \hline \end{array}$$

A 11,500
B 1,150
C 10,500
D 1,050
E None of these

6

$255 \times 6 =$ _____

F 1,500
G 1,230
H 1,530
J 3,680
K None of these

7

$54 \times 32 =$ _____

A 1,638
B 1,628
C 270
D 1,728
E None of these

8

$13\overline{)910}$

F 7
G 70
H 6
J 77
K None of these

9

$6{,}212 \div 2 =$ _____

A 316
B 3,016
C 3,101
D 311
E None of these

10

$6\frac{5}{8} - 2\frac{3}{8} =$ _____

F $\frac{1}{4}$
G $4\frac{1}{8}$
H $4\frac{1}{4}$
J $4\frac{1}{2}$
K None of these

11

$12\overline{)16.8}$

A 14
B 1.4
C 0.14
D 10.4
E None of these

12

$32 - 0.017 =$ _____

F 31.083
G 31.83
H 14
J 30.3
K None of these

13

$$\begin{array}{r} \$12.00 \\ -\ 0.17 \\ \hline \end{array}$$

A \$11.83
B \$11.93
C \$12.93
D \$12.17
E None of these

14

$$\begin{array}{r} \frac{11}{15} \\ +\ \frac{4}{15} \\ \hline \end{array}$$

F $\frac{7}{15}$
G $\frac{1}{2}$
H 1
J 15
K None of these

15

$\frac{3}{4} \div 3 =$ _____

A $\frac{1}{4}$
B $2\frac{1}{4}$
C 4
D $\frac{3}{4}$
E None of these

16

45% of 200 = _____

F 90
G 900
H 18
J 180
K None of these

17

$0.37 \times 3 =$ _____

A 11.1
B 1.11
C 111
D 0.111
E None of these

18

$17 - (-5) =$ _____

F 12
G 22
H –12
J –22
K None of these

19

15% of ☐ = 18

A 27
B 120
C 180
D 150
E None of these

20

$\frac{-36}{-90} =$ _____

F $\frac{2}{5}$
G $-\frac{2}{5}$
H $\frac{4}{9}$
J $-\frac{4}{5}$
K None of these

21

What percent of 150 is 75?

A 75%
B 50%
C 20%
D 12%
E None of these

22

$-5 + (-3) + 2 =$ _____

F –10
G 0
H –6
J 10
K None of these

23

$\frac{3}{5} \times \frac{2}{3} =$ _____

A $\frac{1}{10}$
B $\frac{9}{10}$
C $2\frac{1}{5}$
D $\frac{2}{5}$
E None of these

24

20% of ☐ = 60

F 30
G 120
H 80
J 12
K None of these

25

$5 - (-1) =$ _____

A 4
B –4
C 6
D –6
E None of these

Part B: Applied Mathematics

Circle the letter for the correct answer to each problem.

1 Which of these is another way to show the number 3,010?

A three hundreds, one ten
B $(3 \times 1000) + (1 \times 10)$
C $30 + 10$
D three thousand, one hundred

2 In which of these equations is x equal to 9?

F $x + 9 = 9$
G $9 - x = 0$
H $9 \times x = 18$
J $4 + x = 9$

3 Which of these fractions, if any, is greater than 1?

A $\frac{3}{4}$
B $\frac{12}{15}$
C $\frac{3}{2}$
D None of these

4 Which of these numbers is thirty thousand, four hundred twelve?

F 3,412
G 34,012
H 30,412
J 3,402

5 Which is the total value of three quarters, one nickel, and one dime?

A $0.85
B $85.00
C $0.90
D $1.00

6 One glass goblet costs $14.95. Which of these is the best estimate of how much 21 goblets would cost?

F $300.00
G $210.00
H $280.00
J $200.00

7

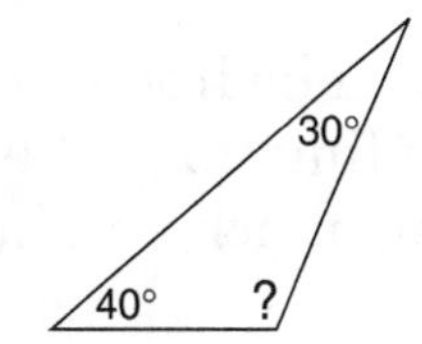

What is the measure of the unlabeled angle in the triangle above?

A 90°
B 100°
C 110°
D There is no way to tell.

8 A box containing 15 marigold plants costs $3.00. How much would 25 plants cost?

F $4.00
G $4.50
H $4.75
J $5.00

This list describes the installment plan at a furniture store. Study the description. Then do Numbers 9 through 12.

> Moore's Furniture Installment Plan
> - 15% down
> - 18% annual interest
> - No payments for 6 months
>
> Minimum monthly payment:
>
> Only $\frac{1}{10}$ of your balance

9 Sean decides to buy a bedroom set for $3,190.00, including tax. If he puts 15% down, how much will he have left to pay?

A $478.50
B $47,850.00
C $2,711.50
D $2,712.50

10 Sean's bedroom set was priced at $2,900.00, and he paid $290 in tax. What tax rate was he charged?

F 29%
G 2.9%
H 10%
J 19%

11 Which of these fractions shows how much interest the store charges each year?

A $\frac{18}{1}$ **C** $\frac{9}{50}$
B $\frac{18}{10}$ **D** $\frac{1}{18}$

12 Which of these shows how long Sean can wait before making his first payment?

F $\frac{1}{6}$ of a year
G $1\frac{1}{2}$ years
H $\frac{2}{3}$ of a year
J $\frac{1}{2}$ of a year

This list shows the ingredients needed to make 4 servings of pancakes. Study the list. Then do Numbers 13 through 16.

> **Pancakes**
>
> $\frac{1}{2}$ cup flour
>
> $\frac{1}{2}$ tsp baking powder
>
> $\frac{1}{4}$ tsp salt
>
> $\frac{1}{2}$ tsp sugar
>
> 1 egg
>
> $\frac{1}{4}$ to $\frac{1}{2}$ cup milk

13 This recipe calls for twice as much sugar as which ingredient?

A baking powder
B salt
C milk
D flour

14 Which of these would be a proper quantity of milk for 4 servings of pancakes?

F $\frac{1}{5}$ cup **H** $\frac{3}{4}$ cup
G $\frac{2}{3}$ cup **J** $\frac{1}{3}$ cup

15 Dan is going to triple the recipe above. How many servings of pancakes will he make?

A 7 **C** 14
B 12 **D** 15

16 How many cups of flour will Dan need to triple the recipe?

F $\frac{1}{6}$
G $1\frac{1}{2}$
H 2
J $\frac{1}{5}$

Mrs. Mahon ordered a hat rack from a catalog. The parts need to be assembled, and the diagram below shows all the pieces that came with her order. Study the diagram. Then do Numbers 17 through 21.

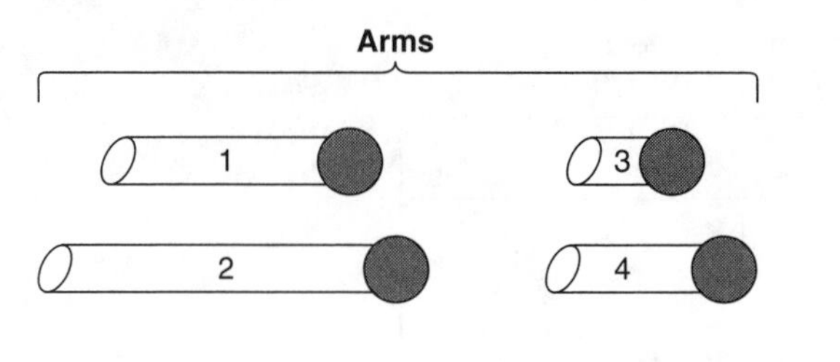

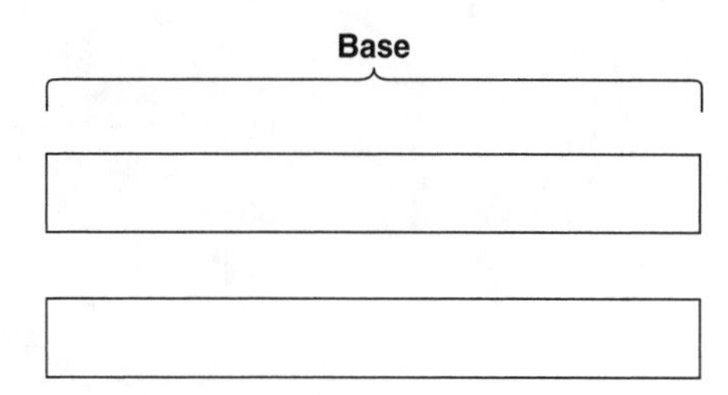

Stand

4 1/2 feet

17 Which two parts of this diagram are congruent?

A Arm 3 and Arm 4
B Arm 1 and Arm 2
C the pieces of the base
D the arms and the stand

18 The directions say that Arm 1 should be $\frac{1}{4}$ of the way down from the top of the stand. How many feet from the top of the stand should Arm 1 be?

F 2 feet
G $1\frac{1}{8}$ feet
H $1\frac{1}{2}$ feet
J $1\frac{1}{4}$ feet

19 What shape is formed by the rounded end of each arm?

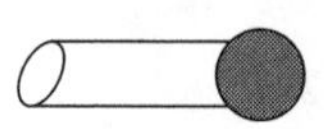

A a sphere
B a parallelogram
C a cone
D a circle

20 The directions say that there should be a 30° angle between each arm and the stand. Which of these shows an angle that is about 30°?

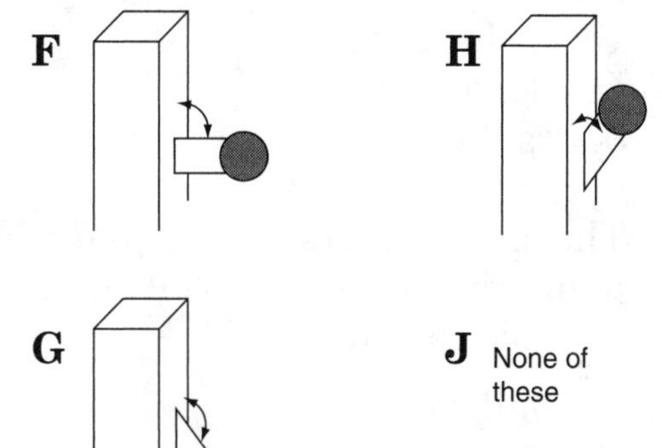

21 The directions say that the two pieces of the base should be nailed together so that one piece is perpendicular to the other. Which of these shows how the base should look?

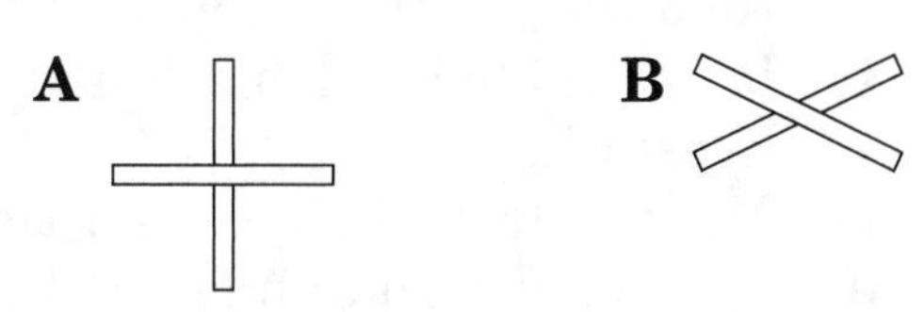

C

D

This graph shows the average number of years different types of pets live. It also shows the longest possible life each type of pet could have. Study the graph. Then do Numbers 22 through 26.

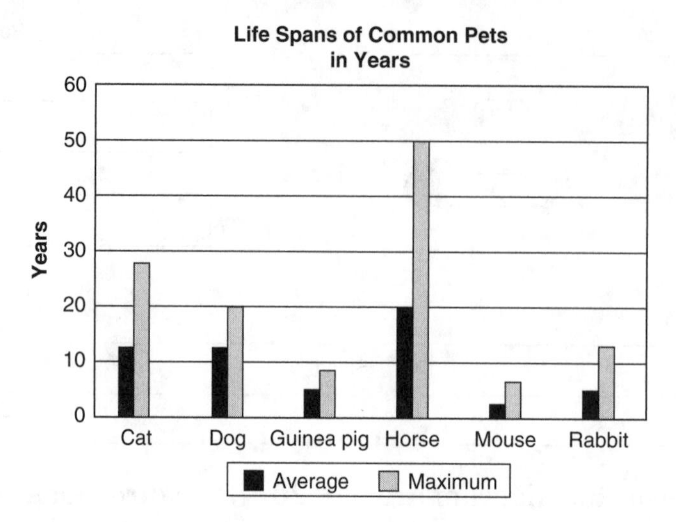

22 Which of the animals listed has the longest average lifespan?

- **F** cats
- **G** Guinea pigs
- **H** horses
- **J** rabbits

23 The oldest of cats live 28 years. How many times longer is that than the average cat's lifespan?

- **A** half as long
- **B** twice as long
- **C** 4 times as long
- **D** 3 times as long

24 Which of the following statements is supported by the information in this graph?

- **F** Animals live longer in the wild than they do in captivity.
- **G** Pets tend to live longer than livestock.
- **H** Many pets die before their time.
- **J** The smaller an animal is, the shorter its lifespan tends to be.

25 Which two pets, if any, have about the same average lifespan?

- **A** cats and dogs
- **B** mice and rabbits
- **C** horses and dogs
- **D** none of them

26 Which animal lives an average life span about half as long as an average horse?

- **F** cat or dog
- **G** mouse
- **H** Guinea pig
- **J** rabbit

The Brookfield Women's Shelter is holding a fashion show to raise money. This diagram shows their stage. Study the diagram. Then do Numbers 27 through 31.

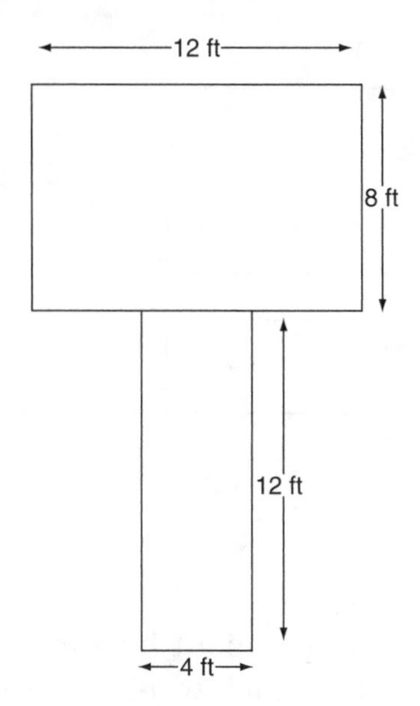

27 How much carpeting would you need to cover the entire stage?

A 72 square feet
B 96 square feet
C 144 square feet
D 48 square feet

28 The steps leading up to the stage will each be 9 inches tall. The stage is 3 feet tall. How many steps will there be leading up to the stage?

F 3
G 4
H 5
J 6

29 There will be cloth tacked around the perimeter of the stage to hide all the supports underneath. How much cloth will be needed?

A 36 feet
B 64 feet
C 68 feet
D 56 feet

30 There will be 28 outfits shown during the show, and each one will take the stage for about 45 seconds. About how long will the show last?

F 126 minutes
G 21 minutes
H 2 hours
J 45 minutes

31 This balance sheet lists all the show's income and expenses. About how much money did the show make? (Round all amounts to the nearest ten dollars.)

Ticket Sales	**+$5,215.00**
Hall Rental	**–$450.00**
Catering	**–$1,625.00**
Decorating	**–$375.95**

A $3,760.00
B $2,760.00
C $7,680.00
D $2,770.00

32 Frank's cable bill is $27.50 per month. Which of these number sentences could be used to find how much he pays for cable in one year?

F $27.50 + 12 = ☐
G $27.50 ÷ 12 = ☐
H $27.50 × 12 = ☐
J $27.50 + $27.50 = ☐

Diane and Jarrad made New Year's resolutions to lose weight. This graph shows how well they did on their diets during the first month. Study the graph. Then do Numbers 33 through 36.

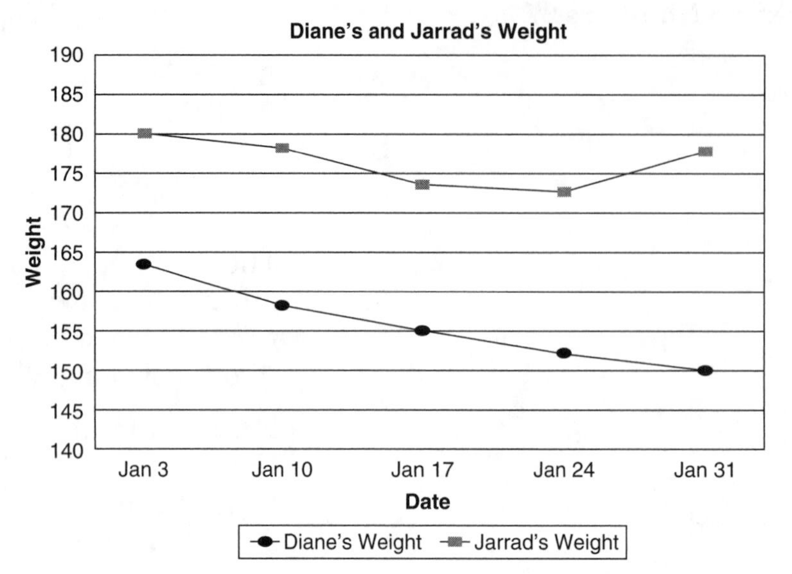

33 In which week did Jarrad lose the most weight?

A January 3 through January 10
B January 10 through January 17
C January 17 through January 24
D January 24 through January 31

34 On which of the dates shown was there the biggest difference between Diane's and Jarrad's weights?

F January 3
G January 10
H January 24
J January 31

35 On average, about how much did Diane lose each week in January?

A $3\frac{1}{2}$ pounds
B 5 pounds
C 6 pounds
D 8 pounds

36 If she continues to lose weight at this rate, how much will Diane weigh at the end of March?

F 163 pounds
G 135 pounds
H 122 pounds
J 100 pounds

Every year the city of Munro hosts a race. This map shows the route that runners take. The dotted line is a shortcut they can take if they do not want to run the full route. Study the map. Then do Numbers 37 through 43.

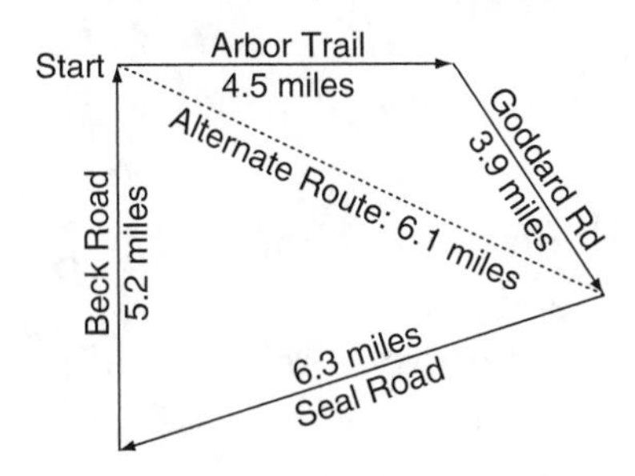

37 How long is the main route?

A 18.9 mi **C** 20.1 mi
B 19.9 mi **D** 20.9 mi

38 On this map, 2 inches equals 10 miles. What scale was used to draw the map?

F 1 inch = 1 mile
G 1 inch = 2 miles
H 1 inch = 5 miles
J 1 inch = 20 miles

39 Sheila took Arbor Trail and Goddard Road but then took the alternate route to the finish line. In which of these equations is x the number of miles by which she shortened the main route?

A $4.5 + 3.9 + 6.1 = x$
B $6.3 + 5.2 = x$
C $(6.3 + 5.2) - 6.1 = x$
D $\frac{6.3 + 5.2}{6.1} = x$

40 The organizing committee decides to put a drink stand every half mile along the route. How many half miles are in the Arbor Trail?

F 2 **H** 5
G 3 **J** 9

41 It took Sheila 47 minutes to run the length of the Arbor Trail. About how fast was she running?

A 5 miles an hour
B 10 miles an hour
C 6 miles an hour
D 4.5 miles an hour

42 At the race, Sheila spent \$15.95 on her entrance fee, \$14.72 on a T-shirt, and \$7.45 on lunch. Which expression gives the *best* estimate of how much she spent altogether?

F \$16.00 + \$15.00 + \$8.00
G \$16.00 + \$15.00 + \$7.00
H \$15.00 + \$14.00 + \$7.00
J \$16.00 + \$14.00 + \$7.00

43 This chart shows the finishing times of the first 10 runners who completed the full route.

Runner	Time
52	3.4 hours
13	3.9 hours
9	4.2 hours
5	4.3 hours
11	4.6 hours
18	4.7 hours
27	4.7 hours
32	4.8 hours
51	5.1 hours
15	5.3 hours

Which of the following is the runner's mean or average finishing time?

A 3.5 hours
B 4.1 hours
C 4.5 hours
D 4.8 hours

44 Round each of the following numbers to the nearest whole number.

1.12	0.89	$1\frac{3}{4}$	$\frac{1}{3}$

How many of the numbers round to 1?

F 1
G 2
H 3
J 4

45 Rena gets 20% off at a clothing store where she works. She bought three scarves that were marked $24.95 each. About how much did she save in all? (Round to the nearest dollar.)

A $5.00
B $6.00
C $15.00
D $18.00

46 The chart below shows how a function changes each "Input" number to a corresponding "Output" number. What number is missing from the table?

Input	5	1	$2\frac{1}{2}$	$5\frac{1}{2}$
Output	$6\frac{1}{2}$	$2\frac{1}{2}$	4	

F 6
G $6\frac{1}{2}$
H $3\frac{1}{2}$
J 7

47 Karin has to print 45 copies of a 50-page report. Paper is sold in 500-sheet packages. To find out how many packages of paper she needs, Karin uses this equation:

$45 \times 50 = n$

What does n represent in this equation?

A how many sheets of paper she needs
B how many packages of paper she needs
C how many reports she can print with one package of paper
D how many sheets of paper are in 4 packages of paper

48 Lee gave $\frac{1}{2}$ a bag of tomatoes to his daughter, $1\frac{1}{2}$ bags to his neighbors, $\frac{3}{4}$ of a bag to the people at work, and $\frac{1}{4}$ of a bag to a golf buddy. How many bags of tomatoes did Lee give away?

F $2\frac{1}{2}$ **H** 2

G $2\frac{3}{4}$ **J** 3

49 The annual city budget of Centerville is 2.5 million dollars. Which of these is the same as 2.5 million?

A $2\frac{1}{2}$ million **C** $2\frac{1}{4}$ million

B 2,005,000 **D** 205 thousand

50 Ron is a fireman. He always works three days in a row, then takes off 4 days in a row. This month, he worked on the 3rd, 4th, and 5th. Which of the following days this month will he be off?

F 10th **H** 14th
G 12th **J** 17th

Skills Inventory Post-Test Evaluation Charts

Use the key to check your answers on the Post-Test. The Evaluation Chart shows where you can turn in the book to find help with the problems you missed.

Keys

Part A

1	A	13	A
2	J	14	H
3	B	15	A
4	G	16	F
5	A	17	B
6	H	18	G
7	D	19	B
8	G	20	F
9	E	21	B
10	H	22	H
11	B	23	D
12	K	24	K
		25	C

Part B

1	B	26	F
2	G	27	C
3	C	28	G
4	H	29	B
5	C	30	G
6	F	31	B
7	C	32	H
8	J	33	B
9	C	34	J
10	H	35	A
11	C	36	H
12	J	37	B
13	B	38	H
14	J	39	C
15	B	40	J
16	G	41	C
17	C	42	G
18	G	43	C
19	A	44	G
20	H	45	C
21	A	46	J
22	H	47	A
23	B	48	J
24	J	49	A
25	A	50	H

Evaluation Chart

Part A

Problem Numbers	Skill Areas	Practice Pages
3, 5, 6, 7	Multiplication of Whole Numbers	14–17
1, 4, 8, 9	Division of Whole Numbers	18–22
2, 11, 12, 13, 17	Decimals	30–41
10, 14, 15, 23	Fractions	42–59
18, 20, 22, 25	Integers	60–66
16, 19, 21, 24	Ratios/ Proportions/ Percent	67–79

Part B

Problem Numbers	Skill Areas	Practice Pages
1, 3, 4, 5, 8, 11, 13, 14, 27, 38, 49	Numeration/ Number Theory	1–5, 30–31, 42–47, 60
22, 23, 24, 25, 26, 33, 34, 35	Data Interpretation	80–93
2, 32, 36, 39, 46, 47, 50	Pre-Algebra/ Algebra	94–105
12, 18, 27, 29, 30	Measurement	106–121
7, 17, 19, 20, 21	Geometry	122–136
9, 10, 15, 16, 28, 37, 40, 48	Computation in Context	23–26, 39, 57, 64
6, 31, 41, 42, 43, 44, 45, 46	Estimation/ Rounding	6–7, 27

Answer Key

Page 1, Place Value

1. ten thousands, thousands, hundreds, tens, and ones
2. 1 hundred thousand, 7 thousands, 4 hundreds, and 3 ones (The student may indicate that there is zero in the remaining places.)
3. 1 million and 5 hundred thousands (The student may indicate that there is zero in the remaining places.)
4. ten thousands
5. hundred thousands
6. 2
7. 5
8. five million(s)
9. six hundred thousand(s)
10. seven thousand(s)
11. three hundred thousand(s)
12. five ten thousands, or fifty thousand

Page 2, Expanded Form
(Places containing the digit 0 may or may not be mentioned.)

1. (5 × 1,000) + (1 × 100) + (0 × 10) + (0 × 1); 5,000 + 100
2. (7 × 10,000) + (5 × 1,000) + (0 × 100) + (0 × 10) + (0 × 1); 70,000 + 5,000
3. (1 × 100,000) + (1 × 10,000) + (2 × 1,000) + (0 × 100) + (2 × 10) + (5 × 1); 100,000 + 10,000 + 2,000 + 20 + 5
4. (4 × 10,000) + (3 × 1,000) + (0 × 100) + (2 × 10) + (0 × 1); 40,000 + 3,000 + 20
5. (5 × 100,000) + (0 × 10,000) + (3 × 1,000) + (0 × 100) + (0 × 10) + (0 × 1); 500,000 + 3,000

Page 3, Naming Large Numbers

1. fifty-nine thousand, five hundred five
2. two hundred eleven thousand, five hundred twenty-five
3. four million, five hundred thousand, seventy-two
4. nine hundred one thousand, three hundred sixteen
5. twenty-three thousand, twelve
6. 91,201
7. 63,412
8. 3,400,000
9. 801,356

Page 4, Comparing Whole Numbers

1. 12,301
2. 946,203
3. 55,000
4 112
5. 91,415
6. 1,000,000
7. 12,000
8. 45,678
9. 1,512
10. 999,999
11. 1,042
12. 899
13. 812
14. 750
15. 1,596
16. 52 115 269 397
17. 952 1,736 6,710 9,000
18. 952 978 5,107 5,123 17,008
19. 9,210
20. 8,651
21. 7,321

Page 5, Writing Dollars and Cents

1.

cents	dollars	pennies	nickels	dimes	quar
100	1	100	20	10	4
50	$\frac{1}{2}$	50	10	5	2
25	$\frac{1}{4}$	25	5	X	1
10	$\frac{1}{10}$	10	2	1	X
5	$\frac{1}{20}$	5	1	X	X

2. 60
3. 50
4. 50
5. 40
6. $1.03
7. $1,046.42
8. five dollars and two cents
9. twelve dollars and sixty cents

Page 6, Estimating

1. not circled
2. not circled
3. not circled
4. circled
5. circled
6. circled
7. circled
8. not circled
9. H
10. A
11. H
12. B
13. H

Page 7, Rounding

1. 180
2. 810
3. 100
4. $11.60
5. $9.20
6. 2,000
7. 20
8. 30
9. 90

10. $1.00
11. 90 cents
12. 200
13. 700
14. 3,100
15. 1,910
16. 25,000
17. 1,000
18. $3.00
19. $12.60
20. $90.00

Pages 8–9, Number System Skills Practice

1. B
2. H
3. D
4. J
5. B
6. H
7. C
8. J
9. B
10. G
11. D
12. F
13. B
14. J
15. A
16. H

Page 10, Review of Whole Number Addition

1. 86
2. $18.67
3. 557
4. 129 min
5. 673 gal
6. 12,893
7. 84,767
8. 1,061
9. 186
10. 861

Page 11, Carrying

1. $35.00
2. 1,138
3. $144.00
4. $19.25
5. 5,660
6. 1,109
7. 878
8. 253
9. 7,408
10. 2,725
11. 911
12. 1,747

Page 12, Review of Whole Number Subtraction

1. 201 miles
2. 331
3. $22.00
4. 53 in.
5. $132.00
6. 1,250
7. 761
8. $12.34
9. 51
10. 112

Page 13, Borrowing

1. 853
2. 196
3. 424
4. $3.89
5. 3,095
6. 2,250
7. $26.20
8. $49.00
9. 819
10. $37.50
11. 537
12. 7,988

Page 14, Multiplying by a One-Digit Number

1. 248
2. 930
3. 2,555
4. 486
5. 808
6. $8.42
7. $14.08
8. 24,408
9. $168.00
10. 1,569
11. 300
12. 620
13. 1,533
14. 264
15. 639

Page 15, Multiplying by a Two-Digit Number

1. 9,360
2. 1,122
3. 1,470
4. 90,630
5. 37,560
6. $23.10
7. 2,626
8. $75.90
9. 300
10. 2,100
11. $29.90
12. $17.60

Pages 16–17, Carrying When You Multiply

1. 125
2. 1,050
3. 18,944
4. 207
5. $87.72
6. 39,585
7. $4.44
8. 336
9. 12,150
10. 5,253
11. $22.95
12. $40.80
13. 14,208
14. 7,875
15. $60.80
16. 11,880
17. $6.75
18. 5,712
19. 4,692
20. 41,250
21. $192.60
22. 14,670
23. $243.78
24. 3,016
25. 1,914
26. 49,665
27. $348.50
28. 90,810
29. 68,733
30. $301.95

Page 18, Using the Division Bracket)‾

1. 124
2. 31
3. 123
4. 42
5. 108
6. 12
7. 314
8. 21
9. 104

Page 19, Dividing with a Remainder

1 13 r 1
2. 10 r 2
3. 11 r 2
4. 2 r 1
5. 10 r 5
6. 4 r 2
7. 2 r 2
8. 21 r 1
9. 102 r 2
10. 103 r 4
11. 203 r 5
12. 304 r 3
13. 109 r 4
14. 104 r 5

Page 20, Writing the Steps of a Division Problem

1. 86
2. 56
3. 161
4. 54
5. 13
6. 112
7. 77
8. 34
9. 221
10. 23
11. 550
12. 38
13. 22
14. 451
15. 131
16. 35

Page 21, Long Division

1. 166
2. 144
3. 133
4. 442
5. 215
6. 3,048

Page 22, Dividing by a Two-Digit Number

1. 20
2. 12
3. 20
4. 32
5. 23
6. 30
7. 4
8. 20
9. 21

Pages 23–24, Solving Word Problems

1. B
2. J
3. A
4. J
5. C
6. G
7. 10, 6, and C should be circled.
8. 50,000; 15,125; and G should be circled.
9. $60.00, 3, and C should be circled.
10. 25, 15, and F should be circled.
11. how much it cost to rent the costume for one day
12. how many secretaries there are
13. the size of Chris' living room

Page 25, Set-Up Problems

1. Divide $120 by 3.
Answer: $40
2. Multiply 23 cents by 20.
Answer: $4.60
3. Subtract 53 from 89.
Answer: 36 boxes
4. Subtract $1,800 and $700 from $3,500.
Answer: $1,000
5. Add $18 and $9.
Answer: $27
6. Multiply 8 by 5, then add 2.
Answer: 42 hours
7. Multiply $18 by 8.
Answer: $144
8. Add $2.00 to $18.00. Multiply the sum by 40.
Answer: $800.00.

Page 26, Solving Two-Step Word Problems

1. B
2. F
3. C
4. 45 minutes
5. $14.75
6. $3,026
7. 11 toys

Page 27, Using Estimation in Word Problems

1. C
2. G
3. B
4. H
5. Divide 180 by 3.
Answer: 60 mph
6. Add $73.00, $125.00, and $120.00.
Answer $318.00
7. Multiply $3.00 by 6.
Answer: $18.00
8. Multiply 20 cents by 30.
Answer: $6.00

Pages 28–29, Computation Skills Practice

1. D
2. F
3. B
4. J
5. E
6. F
7. B
8. G

9. C
10. F
11. C
12. F
13. E
14. J
15. B
16. H
17. B
18. J
19. C

Page 30, Decimals

1. hundredths
2. four tenths or $\frac{4}{10}$
3. nine hundredths or $\frac{9}{100}$
4. ten-thousandth
5. thousandths
6. hundredths

Page 31, Comparing Decimal Numbers

1. 100
2. 10
3. 10
4. 1
5. 0.0283
6. 0.0089
7. 0.059
8. 0.005
9. 1.013
10. 1.23
11. 0.01
12. 3.0
13. 12.20
14. 5.010

Page 32, Adding Decimals

1. 32.01
2. 0.23
3. 1.523
4. 1.59
5. 0.105
6. 1.425
7. 7.4
8. 9.23
9. 23.209
10. 12.231

Page 33, Subtracting Decimals

1. 3
2. 0.762
3. 1.485
4. 0.43
5. 3.24
6. 15.575
7. $31.12
8. $0.68
9. 0.234
10. 8.2
11. $15.58
12. 4.983

Pages 34–35, Multiplying Decimals

1. 0.28
2. 0.21
3. 7.5
4. 0.603
5. 8.48
6. 0.1104
7. 1.2
8. 1.12
9. 14.4
10. 0.3
11. 0.105
12. 0.4
13. 14
14. Multiply 0.5 by $2.50.
    Answer: $1.25
15. Multiply 3 dollars by 0.5.
    Answer: $1.50
16. 0.06
17. 0.012
18. 0.00125
19. 0.038
20. 0.00186
21. 0.0035
22. 0.00048
23. 0.36
24. Multiply 0.02 and 12.06.
    Answer: 0.2412
25. Multiply 0.03 and 0.2.
    Answer: 0.006
26. Multiply 0.1 and 0.5.
    Answer: 0.05
27. Multiply 0.3 and 0.09.
    Answer: 0.027
28. Multiply 0.25 and 40.
    Answer: 10
29. Multiply 1.2 and 0.1.
    Answer: 0.12 oz
30. Multiply 0.25 and 60.
    Answer: 15
31. 0.75
32. $1.28
33. $4.80

Page 36, Dividing a Decimal by a Whole Number

1. 2.04
2. 0.063
3. 16.1
4. $2.04
5. 30.09
6. 0.105
7. 0.5
8. 1.023
9. 5.3
10. 0.304
11. 0.27
12. 0.0081
13. 0.126
14. $7.14
15. $36.00

Page 37, Dividing a Decimal by a Decimal

1. 96
2. 6
3. 9
4. 49.6
5. 4.8
6. 24
7. 2.3
8. $26.30
9. 770
10. 714
11. 4 hot dogs
12. 24 minutes

Page 38, Dividing a Whole Number by a Decimal

1. 400
2. 325
3. 60

4. 12,000
5. 6,300
6. 2,700
7. 320
8. 2,870
9. 120
10. 2,400
11. 15 gallons
12. 40 cents

Page 39, Solving Mixed Word Problems

1. D
2. H
3. B
4. F
5. 558.2 miles
6. 3.9 miles an hour
7. $4.94
8. 2 cents

Pages 40–41, Decimals Skills Practice

1. C
2. G
3. D
4. F
5. C
6. K
7. B
8. G
9. D
10. F
11. B
12. J
13. C
14. F
15. A
16. J
17. B
18. J
19. D
20. G
21. B
22. J

Page 42, Numerators and Denominators

1. $\frac{17}{100}$
2. $\frac{5}{12}$
3. $\frac{23}{59}$
4. $\frac{3}{8}$
5. $\frac{12}{8}$ *or* $\frac{3}{2}$
6. 400 ÷ 4 *or* 100
7. 39 ÷ 3 *or* 13
8. 100 ÷ 5 *or* 20
9. 36 ÷ 6 *or* 6
10. 40 ÷ 8 *or* 5
11. 300 ÷ 2 *or* 150

Page 43, Comparing Fractions

1. >
2. <
3. >
4. $\frac{1}{9}, \frac{1}{7}, \frac{1}{5}$
5. $\frac{2}{9}, \frac{2}{5}, \frac{2}{3}$
6. $\frac{1}{9}, \frac{1}{8}, \frac{1}{7}$
7. $\frac{1}{7}, \frac{3}{7}, \frac{5}{7}$
8. $\frac{1}{5}, \frac{2}{5}, \frac{2}{3}$
9. $\frac{2}{9}, \frac{4}{9}, \frac{4}{5}$
10. $\frac{1}{6}, \frac{1}{4}, \frac{1}{3}$

Pages 44–45, Reducing a Fraction to Simplest Terms

1. $\frac{1}{5}$
2. $\frac{3}{10}$
3. $\frac{2}{5}$
4. $\frac{1}{2}$
5. $\frac{1}{4}$
6. $\frac{3}{10}$
7. $\frac{4}{9}$
8. $\frac{1}{5}$
9. $\frac{1}{32}$
10. $\frac{1}{7}$
11. $\frac{1}{5}$
12. $\frac{1}{4}$
13. $\frac{2}{5}$
14. $24.40
15. $\frac{1}{2}$
16. $\frac{1}{4}$
17. $40.00 off
18. $21.00
19. $\frac{1}{3}$
20. $21.00
21. 6 gal
22. by $4.00
23. $80.00
24. Tony

Page 46, Fractions Equal To 1 and Fractions Greater Than 1

1. <
2. >
3. <
4. <
5. =
6. <
7. =
8. >
9. >
10. >
11. $1\frac{1}{3}$
12. 2
13. $1\frac{1}{4}$
14. 5
15. $2\frac{3}{8}$

16. $1\frac{1}{2}$

17. $1\frac{2}{5}$

18. $1\frac{1}{5}$

19. $1\frac{4}{5}$

20. $3\frac{3}{5}$

Page 47, Changing Between Decimals and Fractions

1. $\frac{2}{5}$
2. $\frac{9}{20}$
3. $4\frac{1}{5}$
4. $\frac{2}{25}$
5. $\frac{4}{5}$
6. 0.2
7. 0.05
8. 0.75

Page 48, Adding Fractions and Mixed Numbers

1. $\frac{7}{9}$
2. 1
3. $1\frac{1}{3}$
4. $\frac{2}{3}$
5. $\frac{1}{2}$
6. $\frac{2}{5}$
7. $\frac{1}{3}$
8. 6
9. $\frac{7}{8}$
10. $1\frac{1}{2}$
11. $2\frac{1}{2}$
12. $\frac{5}{9}$
13. $1\frac{1}{2}$
14. $5\frac{2}{3}$
15. $6\frac{4}{5}$

Page 49, Subtracting Fractions and Mixed Numbers

1. $\frac{1}{3}$
2. $\frac{1}{12}$
3. $1\frac{2}{7}$
4. $\frac{1}{2}$
5. $1\frac{1}{5}$
6. $\frac{2}{3}$
7. $\frac{3}{5}$
8. 2
9. $3\frac{1}{3}$
10. $\frac{3}{4}$
11. $4\frac{1}{4}$
12. $4\frac{2}{5}$
13. 1

Page 50, Borrowing To Subtract Mixed Numbers

1. $\frac{2}{5}$
2. $1\frac{1}{2}$
3. $2\frac{5}{7}$
4. $1\frac{7}{8}$
5. $1\frac{2}{3}$
6. $2\frac{1}{3}$
7. $8\frac{3}{5}$
8. $\frac{13}{15}$
9. $\frac{3}{4}$
10. $1\frac{4}{5}$
11. $4\frac{4}{5}$

Pages 51–52, Adding and Subtracting Unlike Fractions

1. 3
2. 6
3. 12
4. 2
5. 6
6. 9
7. 6
8. 7
9. 5
10. 4
11. 5
12. 6
13. 10
14. 15
15. 18
16. 14
17. 12
18. 45
19. 21
20. $\frac{4}{12}$ and $\frac{3}{12}$
21. $\frac{3}{15}$ and $\frac{1}{15}$
22. $\frac{5}{20}$ and $\frac{4}{20}$
23. $\frac{14}{21}$ and $\frac{6}{21}$
24. $\frac{4}{6}$ and $\frac{5}{6}$
25. $\frac{20}{24}$ and $\frac{15}{24}$
26. $\frac{21}{35}$ and $\frac{10}{35}$
27. $\frac{7}{12}$

28. $\frac{1}{2}$

29. $\frac{5}{6}$

30. $\frac{1}{5}$

31. $\frac{1}{4}$

32. $\frac{5}{9}$

33. $\frac{13}{15}$

34. $\frac{7}{18}$

35. $\frac{5}{12}$

36. $\frac{1}{2}$

37. $\frac{1}{8}$

38. $\frac{13}{24}$

Page 53, Multiplying Fractions

1. $\frac{3}{20}$

2. $\frac{2}{15}$

3. $\frac{4}{9}$

4. $\frac{3}{25}$

5. 3

6. $\frac{2}{5}$

7. $\frac{3}{8}$

8. $\frac{4}{45}$

9. $1\frac{2}{5}$

10. $\frac{1}{6}$

11. $\frac{3}{20}$

12. $\frac{9}{35}$

13. $1\frac{1}{9}$

14. 8

15. 10

16. $\frac{1}{8}$ cup

17. 10 people

Page 54, Canceling Before You Multiply

1. $\frac{2}{15}$

2. $\frac{1}{9}$

3. $\frac{1}{6}$

4. $\frac{1}{30}$

5. $\frac{2}{5}$

6. $\frac{3}{5}$

7. $\frac{1}{5}$

8. $\frac{1}{7}$

9. 11

10. $\frac{1}{2}$ pound

11. $\frac{2}{5}$

Page 55, Dividing a Fraction by a Fraction

1. $1\frac{1}{2}$

2. $1\frac{1}{2}$

3. 4

4. $\frac{2}{3}$

5. $\frac{8}{15}$

6. $2\frac{2}{5}$

7. 2 strips

8. $\frac{4}{5}$

9. 3

10. 6 pitchers

Page 56, Dividing with Fractions and Whole Numbers

1. 18

2. 8

3. 12

4. $\frac{1}{7}$

5. $\frac{1}{6}$

6. $\frac{2}{45}$

7. $\frac{1}{15}$

8. $\frac{1}{21}$

9. 10 sections

10. 8 portions

Page 57, Solving Mixed Word Problems

1. C

2. J

3. A

4. J

5. D

6. yes

7. $\frac{3}{14}$

8. $\frac{5}{6}$ of a tank

9. 8 cups

10. $24.00

Pages 58–59, Fractions Skills Practice

1. B

2. J

3. C

4. G

5. C

6. H

7. A

8. F

9. D

10. F

11. D

12. H

13. C

14. G
15. A
16. H
17. C
18. G
19. C

Page 60, Positive and Negative Numbers

1. <
2. <
3. >
4. <
5. <
6. >
7. 3.1
8. 3
9. 5
10. 5
11. 6
12. 0.25
13. =
14. <
15. <

Page 61, Adding Signed Numbers

1. 3
2. –3
3. 4
4. –1
5. –5
6. –2
7. 0
8. –2
9. –700 ft
10. –4°F
11. –5
12. left

Page 62, Subtracting Signed Numbers

1. 5
2. –5
3. 1
4. –3
5. 6
6. 7
7. –5
8. –10
9. 9
10. 13
11. 10
12. –6
13. –6
14. –3
15. 12,242 feet
16. +$60.00

Page 63, Multiplying and Dividing Signed Numbers

1. –6
2. 20
3. –12
4. –32
5. –30
6. 64
7. 35
8. –18
9. –24
10. –4
11. –8
12. 10
13. –2
14. –2
15. $\frac{1}{3}$
16. $-\frac{1}{2}$
17. $-\frac{2}{3}$
18. $\frac{1}{3}$

Page 64, Solving Mixed Word Problems

1. –$72.41
2. –6 degrees
3. 4 ft per min
4. $19,500
5. 10 minutes
6. $168.00

Pages 65–66, Signed Numbers Skills Practice

1. B
2. J
3. A
4. F
5. A
6. H
7. A
8. H
9. D
10. J
11. B
12. H
13. C
14. H
15. A
16. G
17. D
18. J
19. B

Page 67, Writing Ratios

1. $\frac{\text{1 inch}}{\text{25 miles}}$
2. $\frac{\$550}{\$2958}$
3. $\frac{\text{1 evergreen}}{\text{3 leafy trees}}$
4. $\frac{\$4.00}{\text{12 roses}}$
5. $\frac{\text{3 shirts}}{\text{price of 1}}$
6. $\frac{\text{3 hours}}{\text{4 placements}}$
7. $\frac{\text{450 miles}}{\text{1 hour}}$
8. $\frac{\text{3 men}}{\text{2 women}}$
9. $\frac{\text{3 ears}}{\$1.00}$
10. $\frac{\$17\text{ in materials}}{\$60\text{ in labor}}$
11. $\frac{\text{3 losses}}{\text{13 games}}$
12. $\frac{\text{12 days work}}{\text{5 days off}}$

Page 68, Finding a Unit Rate

1. $3.00
2. $16.00
3. $5.00
4. $1,333
5. 12 or 13
6. 3.5
7. Savory Soup

Pages 69–70, Writing Proportions

(These proportions may be inverted, as long as the ratios on both sides are inverted.)

1. $\frac{\text{12 yards}}{\text{\$40.00}} = \frac{\text{36 yards}}{\text{\$120.00}}$
2. $\frac{\text{\$50.00}}{\text{1 hour}} = \frac{\text{\$150.00}}{\text{3 hours}}$
3. $\frac{\text{2 pounds}}{\text{\$1.00}} = \frac{\text{1 pound}}{\text{\$0.50}}$
4. $\frac{\text{1 inch}}{\text{3 feet}} = \frac{\text{3 inches}}{\text{? feet}}$
5. $\frac{\text{15 feet}}{\text{1 day}} = \frac{\text{60 feet}}{\text{? days}}$
6. $\frac{\text{1 piece}}{\text{500 calories}} = \frac{\text{3 pieces}}{\text{? calories}}$
7. 3
8. 20
9. 12
10. 2
11. 10
12. 240
13. 10
14. 10
15. 130
16. 5 hours
17. $25.00
18. 18 sales
19. 1,050 miles
20. 8 cans
21. 25 weeks

Page 71, Cross Multiplying to Solve a Proportion

1. A
2. G
3. 25
4. 15
5. 12

Page 72, Percent

Percent	Fraction	Decimal
5%	$\frac{1}{20}$	0.05
8%	$\frac{2}{25}$	0.08
10%	$\frac{1}{10}$	0.1
12%	$\frac{3}{25}$	0.12
15%	$\frac{3}{20}$	0.15
18%	$\frac{9}{50}$	0.18
20%	$\frac{1}{5}$	0.2
25%	$\frac{1}{4}$	0.25
30%	$\frac{3}{10}$	0.3
40%	$\frac{2}{5}$	0.4
50%	$\frac{1}{2}$	0.5
55%	$\frac{11}{20}$	0.55
60%	$\frac{3}{5}$	0.6
72%	$\frac{18}{25}$	0.72
75%	$\frac{3}{4}$	0.75
78%	$\frac{39}{50}$	0.78
80%	$\frac{4}{5}$	0.8
85%	$\frac{17}{20}$	0.85
90%	$\frac{9}{10}$	0.9

Page 73, Finding a Percent of a Number

1. 21
2. $6.30
3. $4.50
4. $690.00
5. 120
6. $2.58

Page 74, Adding or Subtracting Percent

1. $40.00
2. $74.75
3. $3.78
4. 6,775 votes
5. $25,200
6. $54.00

Page 75, Finding What Percent One Number Is of Another

1. 20%
2. 60%
3. 13%
4. 40%
5. 4%
6. 40%

Page 76, Finding the Total When a Percent Is Given

1. 20,960
2. $15,100
3. $500
4. $60,000

Page 77, Mixed Practice with Percent

1. 57%
2. $25.60
3. $55.00
4. 1,501 votes
5. $19.60
6. 22%
7. $6.75
8. 70%

Pages 78–79, Ratio and Percent Skills Practice

1. C
2. F
3. C
4. J
5. B
6. J
7. B
8. J
9. B
10. F
11. D
12. J
13. D
14. G
15. B
16. F

Page 80, Reading a Table
1. B
2. November 21, 1980
3. 50,150,000
4. Roots, Dallas, and M*A*S*H
5. The last episode of M*A*S*H
6. Roots, Part 8

Page 81, Using Numbers in a Table
1. The Cherokee
2. The Choctaw
3. 5,016
4. The Sioux
5. The Navajo
6. D
7. 30,901
8. The Sioux or the Chippewa

10. No

Page 82, Using a Price List
1. \$3.00
2. \$2.08
3. \$6.41
4. \$7.66
5. \$12.00
6. \$50–\$100
7. \$3.06
8. \$2.87
9. \$1.15
10. \$11.47

Page 83, Finding the Mean, Median, and Mode

Median	Mode	Mean
179.5	None	181
180	None	180
230	230	227
173.5	175	172
165	None	165
121.5	144	121
229	246	214

Page 84, Graphs
1. 8
2. 4
3. 7

Page 85, Reading a Circle Graph
1. networks
2. independent stations
3. independent stations
4. independent stations
5. $\frac{36}{100}$ or $\frac{9}{25}$
6. 1988
7. basic cable and pay cable

Page 86, Numbers and Percents in a Circle Graph
1. 621
2. 399
3. 1,065
4. bumped from flight

Page 87, Reading a Bar Graph
1. 500 million
2. C
3. two times
4. C

Page 88, Using a Key or Legend
1. 200
2. 500 out of every 100,000
3. heart disease and cancer
4. about $\frac{1}{5}$
5. heart disease
6. about 150 deaths per 100,000
7. B
8. about 400
9. heart disease
10. heart disease

Page 89, Reading a Line Graph
1. fall
2. 1984–1986
3. 1986
4. 2 million
5. a little over 2 million
6. about $5\frac{1}{2}$ million
7. 1996

Pages 90–91, Trends and Predictions
1. B
2. H
3. B
4. H
5. A
6. F
7. C

Pages 92–93, Data Interpretation Skills Practice
1. C
2. F
3. B
4. H
5. D
6. G
7. D
8. J
9. C
10. J
11. A

Page 94, Patterns

1. 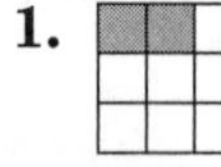

2. 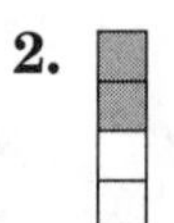

3. 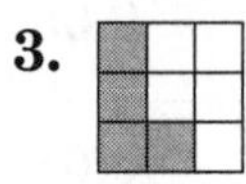

4.

5.  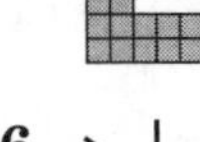

6. 

7. 

8. 

9. 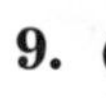

10. 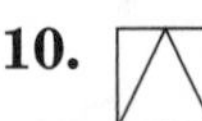

Page 95, Finding Number Patterns

1. even
2. 2
3. 3
4. 25
5. 14, 17, 20
6. 19, 35, 67
7. 3
8. minus 6
9. divided by two (or halved)
10. minus 15
11. plus 5
12. plus 3
13. plus 12
14. 2

Page 96, Patterns in Number Sentences

1. +
2. –
3. +
4. –
5. ×
6. –
7. –
8. ×
9. –
10. ÷
11. ×
12. ÷
13. ×
14. ÷
15. ×
16. 3
17. 2
18. 5
19. 20
20. 8
21. 30
22. 10
23. 10
24. 15
25. 32
26. 12
27. 2
28. 11
29. 10

Page 97, Some Basic Number Properties

1. ÷
2. –
3. × or ÷
4. – or +
5. ×
6. 915
7. –35
8. –692
9. 44, true
10. 56, true

Page 98, Functions

1. Add 6.
2. Subtract 25.
3. Multiply by 3.
4. 9
5. 10
6. 18
7. F
8. C

Page 99, Writing Letters and Symbols for Words

(The order of the numbers in any of these multiplication or addition problems may be reversed)

1. $16x$
2. $12 + x$
3. $5x$
4. $x + 17$
5. $x - 6$
6. $x + 10$
7. $x \div 2$ *or* $\frac{x}{2}$
8. $|20 - x|$ *or* $|x - 20|$
9. $\frac{1}{8}x$ *or* $\frac{x}{8}$ *or* $x \div 8$
10. $\frac{x}{3}$ *or* $x \div 3$
11. $x + 5$
12. $x - 2$
13. $2x$
14. $5x$
15. $x + 6$
16. $5 \div x$ or $\frac{5}{x}$

Page 100, Writing Two-Step Algebraic Expressions

1. $2n - 12$
2. $\frac{n}{4} + 6$
3. $2n + 15$
4. $\frac{n}{3} + 2$
5. $(n + 5) - 1$

Page 101, Using Algebra to Solve Word Problems

1. $d + 15$
2. $h \div 4$ *or* $\frac{\text{h}}{4}$
3. $8 - x$
4. $6c$
5. $t - 6$
6. $4m + 30$
7. C
8. F
9. C

Pages 102–103, Solving Equations

1. $x = 18$
2. $x = 52$
3. $x = 6$
4. $x = 88$
5. $x = 61$
6. $x = 75$
7. $x = 25$
8. $x = 16$
9. $x = 20$
10. $n = 16$
11. $x = 15$
12. $a = 12$
13. $n = 36$
14. $b = 12$

15. $a = 2$
16. $x = 66$
17. $x = 50$
18. $g = 25$
19. $x = 3$
20. $n = 20$
21. $b = 150$
22. $t = 20$
23. 5
24. 24
25. 82
26. 14
27. 15

Pages 104–105, Algebra Skills Practice

1. C
2. F
3. A
4. H
5. D
6. G
7. D
8. G
9. A
10. J
11. C
12. J
13. B
14. J

Pages 106–107, Reading a Scale

1. 19
2. 95
3. 70
4. 65
5. 25
6. 70
7. 45
8. 75
9. 2
10. 2
11. 5
12. 10
13. 20
14. 4
15. 25
16. 2, 32
17. 10, 20

Pages 108–109, Using a Ruler

1. $1\frac{1}{2}$
2. 2
3. $\frac{1}{2}$
4. $2\frac{1}{2}$
5. 1
6. $\frac{1}{4}$
7. $\frac{3}{4}$
8. $1\frac{1}{4}$
9. $\frac{1}{2}$
10. $1\frac{3}{4}$
11. $\frac{1}{2}$
12. 1
13. $\frac{1}{2}$
14. $\frac{1}{2}$
15. 1
16. D
17. G
18. $\frac{1}{2}$
19. $\frac{3}{4}$
20. $1\frac{1}{2}$
21. $\frac{7}{8}$
22. $\frac{1}{4}$
23. 3
24. $\frac{15}{16}$
25. $\frac{7}{16}$
26. $\frac{11}{16}$
27. $\frac{9}{16}$
28. $\frac{13}{16}$

Page 110, Measuring Real Objects

1. The student must measure *two* of the following objects.
   cassette tape: 4 by $2\frac{1}{2}$ (or $2\frac{2}{4}$)
   new pencil: $7\frac{1}{2}$ by $\frac{1}{4}$
   index card: 3 by 5
   paper bag: 17 by $11\frac{3}{4}$ or 17 by 7
2. The student must measure *two* of the following objects.
   can of soda: 5 by $2\frac{1}{2}$
   1-gal milk jug: 10 by 6
   12-oz juice can: 5 by $2\frac{1}{2}$
   2-liter soda bottle: 12 by $4\frac{1}{2}$
3. The student must measure *two* of the following objects.
   Campbell's® soup can: $3\frac{7}{8}$ by $2\frac{5}{8}$
   postage stamp: 1 by $\frac{5}{8}$
   computer disk: $3\frac{1}{2}$ by $3\frac{5}{8}$
   a dollar bill: $6\frac{1}{8}$ by $2\frac{5}{8}$
   a 9-volt battery: $1\frac{7}{8}$ by 1, or $1\frac{7}{8}$ by $\frac{5}{8}$
4. A
5. H
6. D

Page 111, Standard Units of Measure

1. C
2. G
3. G
4. A
5. 15 inches

6. 2 yards
7. 1 pound
8. 22 ounces
9. 3 cups
10. 1 quart
11. 1 gallon

Page 112, Converting Within the Standard System

1. 3
2. 2
3. 1
4. 6
5. 1
6. 18
7. 5
8. 14
9. 8
10. 30
11. 41
12. 3, 2
13. 2, 2
14. 48
15. 8
16. 1, 7
17. 35
18. 6
19. 90
20. 15

Page 113, The Metric System

1. kilogram
2. liter
3. milliliter
4. centimeters
5. meter
6. kilograms
7. 500
8. 2,000
9. 3
10. 1,500
11. 3.5
12. 100
13. 2,500
14. 1.5
15. 315
16. 1,450
17. 3,015
18. 1, 515

Page 114, Comparing Standard and Metric Units

1. 1 inch
2. 1 ounce
3. 1 liter
4. 1 meter
5. 1 mile
6. 18.2
7. 1.4
8. 25
9. 11
10. 3.3
11. 6.8
12. 40
13. 0.5
14. 30

Page 115, Adding and Subtracting Mixed Measurements

1. 4 hours
2. 10 pounds 1 ounce
3. 8 yards 1 foot
4. 16 feet 5 inches
5. 2 hours 5 minutes
6. 3 quarts 1 cup
7. 7 pounds 3 ounces
8. 1 yard 1 foot
9. 3 feet 8 inches
10. 2 quarts or $\frac{1}{2}$ gallon
11. 2 hours 30 minutes
12. 2 pints 1 cup

Page 116, Finding Perimeter

1. 55 feet
2. 22' 10"
3. 48 m
4. 16 mm
5. 24 in.
6. 32 km
7. 4 inches

Page 117, Finding Area (Answers must be given in square units)

1. 25 meters^2
2. 150 feet^2
3. 0.25 square yards
4. 84 square miles
5. 90 square inches
6. 3 square feet
7. 3,249 mm^2
8. 24 in.^{2}
9. 495 m^2

Page 118, Finding Volume (Answers must be given in cubic units)

1. 98 in.^{3}
2. 72 in.^{3}
3. 8 m^3
4. 90 cm^3
5. $\frac{1}{8}$
6. 27 inches^3
7. C
8. 12 times

Page 119, Changes in Time

1. 10:15
2. 8:45
3. 4:45
4. 40 minutes
5. 9:10

Pages 120–121, Measurement Skills Practice

1. C
2. G
3. B
4. H
5. B
6. J
7. A
8. J
9. A
10. H
11. C
12. J
13. C

Page 122, Basic Concepts

1. • C
2. D •——• E
3. ←—•S——•T—→

4. 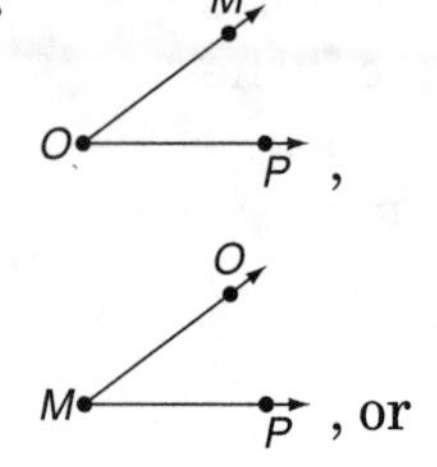

, or

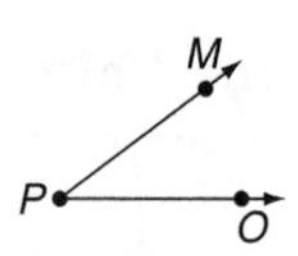

5. angle
6. line
7. A line segment has end points. It does not go on forever.
8. *F, G, H,* and *J*
9. any two of: ∠*FHG*, ∠*GHF*; ∠*FHJ*, ∠*JHF*; ∠*JHG*, ∠*GHJ*; ∠*JGF*, ∠*HGF*, ∠*FGH*, ∠*FGJ*; ∠*HFG*, ∠*GFH*

Page 123, Angles

1. A
2. G
3. A
4. G
5. C
6. F
7. A

Page 124, Types of Angles

1. right
2. acute
3. obtuse
4. right
5. acute
6. obtuse
7. right
8. acute
9. obtuse
10. acute
11. right
12. ∠2 and ∠4
13. B

Page 125, Lines

1. parallel
2. intersecting
3. intersecting
4. parallel
5. yes
6. no
7. no
8. yes
9. $\overleftrightarrow{EF}$ *or* $\overleftrightarrow{FE}$
10. no
11. No parallel lines are shown.

Page 126, Polygons

1. C
2. F
3. C
4. F
5. C
6. G
7. B
8. H
9. B
10. F

Pages 127–128, Types of Triangles

1. isosceles, 50°
2. scalene, 50°
3. equilateral, 60°
4. scalene, 48°
5. right, 45°
6. obtuse, 30°
7. acute, 65°
8. right, 90°
9. A
10. H
11. B
12. H

Page 129, Circles

1. B
2. diameter
3. radius: 2 in.
   diameter: 4 in.
4. radius: 3 mm
   diameter: 6 mm
5. radius: cannot tell
   diameter: cannot tell

Page 130, Finding the Circumference of a Circle

1. 18.84 feet
2. 9.42 meters
3. 47.1 inches
4. 25 inches
5. 7.85 yards
6. 251.2 centimeters

Pages 131–132, Recognizing Congruent and Similar Figures

1. The student should have drawn a figure exactly like the black pentagon next to it. The drawing may be turned a different way, but it must be the same shape and size.
2. The student should have drawn a square, 2 units wide and 2 units tall.
3. C
4. G
5. B
6. G
7. ∠*D* and ∠*C*
8. A and E or C and D
9. segment *DE*
10. A and B
11. C and D

Page 133, Three-Dimensional Figures

1. cylinder
2. cone
3. cone
4. cylinder
5. sphere
6. rectangular solid
7. no (Its sides do not all meet at right angles.)
8. A

Page 134, Visualizing Solids

1. 27
2. D
3. H
4. B
5. F
6. A

Pages 135–136, Geometry Skills Practice

1. A
2. H
3. C
4. G
5. D
6. H
7. B
8. H
9. B
10. J
11. A
12. J

Glossary of Common Terms

absolute value: a number's distance from zero. The symbol for absolute value is | |. Examples: 5 and -5 both have an absolute value of 5.

angle: the figure formed by two lines coming from the same point

area: an amount of a flat region

average: the sum of all the values in a set divided by the total number of values. Example: To find the average of the set 1, 5, 4, 6, 5, and 3, add all six numbers. Then divide the sum (24) by 6. The average is 4.

canceling: a way to simplify multiplication problems that contain fractions. To cancel, use the same number to divide one of the numerators and one of the denominators in the fractions being multiplied.

circumference: the distance around the outside of a circle. To find the circumference of any circle, multiply its diameter by π, which is about 3.14.

common denominator of two fractions: a number that can be evenly divided by the denominators of both fractions. The **least common denominator** is the smallest number that can be evenly divided by both denominators.

congruent figures: figures with the same size and shape

cross multiply: one way to find an unknown number in two equivalent fractions or a proportion. The numerator in one fraction times the denominator in the other fraction equals the product of the two remaining numbers. Example:
If = $\frac{3}{10} = \frac{b}{15}$, then $3 \times 15 = b \times 10$.

decimal: a number containing digits to the right of the **decimal point.** Example: In $13.42, digits to the left of the decimal point represent 13 whole dollars while digits to the right represent 42 hundredths of a dollar.

denominator: the bottom number in a fraction. The denominator tells how many parts are in one unit.

diameter: the width of a circle. The diameter of a circle always passes through its center.

dividend: the number in a division problem that is being divided

divisor: the number in a division problem that you are dividing by

equation: a number sentence showing that numbers or mathematical expressions are equal. Example: $x + 7 = 8$

equivalent: having the same value. Examples: $\frac{1}{2}$, $\frac{4}{8}$, and $\frac{5}{10}$ are all equivalent fractions.

fraction: a way of representing part of something. The bottom number in a fraction tells how many parts there are in the whole. The top number shows how many parts the fraction represents.

function: a rule that changes one number into another number

inverse operations: operations that undo each other. Addition and subtraction are inverse operations, as are multiplication and division. Example: If 12 has been added to a number, you can find the original number by subtracting 12.

key (or legend): an explanation of the symbols used in a map, graph, diagram, or table

line segment: part of a line. A line segment has two endpoints.

lowest terms: *See* simplest terms.

mean: *See* average.

median: the middle value in a set of numbers listed in order, or the number halfway between the two middle values.

mixed number: a combination of a whole number and a fraction. Mixed numbers represent a value between two whole numbers. Example: $2\frac{1}{2}$

mode: the number in a set that appears most often

negative number: a number with a value less than zero. Negative numbers are always shown with a negative sign (–).

numerator: the top number in a fraction. The numerator tells how many parts are represented.

operation: something done to a number, usually addition, subtraction, multiplication, or division

percent: a fraction with an unwritten denominator of 100. The word percent means "per hundred." Example: If 100 people vote in an election and 14 of them vote for Smith, then Smith got 14 percent (14%) of the votes.

perimeter: the total length of the sides in a figure

pi (π): approximately 3.14 or $\frac{22}{7}$

place: the location of a digit in a number. Example: 173 has three places. The digit 1 is in the hundreds place, 7 is in the tens place, and 3 is in the ones place.

point: a single location in space

polygon: a flat shape that has straight sides. Examples: rectangle, hexagon, pentagon. In a **regular polygon,** like a square, all sides have the same length.

proportion: an equation showing that one ratio equals another. Example: = $\frac{3}{4} = \frac{6}{8}$

radius: the distance from a point on a circle to the center of the circle

ratio: a comparison of two numbers. Example: Ron has 3 days off for every 7 days he works. Ratios can be written in three ways: 3 to 7, 3 : 7, or $\frac{3}{7}$.

reduce (a fraction): change into an equivalent fraction with smaller numbers. To reduce a fraction, divide both the top and bottom numbers by the same number.

right angle: a 90-degree angle. Each corner in a square is a right angle.

rounding: increasing or decreasing a number so that it ends in one or more zeros. Example: 209 rounded to the nearest hundred is 200.

scale: a number line on a measurement tool

similar figures: figures with the same shape, but not necessarily the same size

simplest terms: a fraction or a ratio is in simplest terms when there is no whole number, other than 1, that can evenly divide both numbers in it.

three-dimensional (or solid) figure: a figure that has depth as well has length and width. Examples: cube, cone, cylinder.

triangle: a three-sided figure

unknown: a letter (or variable) in an algebraic expression. Unknowns can take the place of numbers. Example: In the expression $2a + 3 = 11$, a is the unknown.

unit rate: the rate for one unit of a given quantity. Examples: 35 miles per gallon or 3 cents per ounce

variable: *See* unknown.

volume: the amount of space within a three-dimensional figure.

Index